Knowledge as a Tale

Rafal Maciag

Knowledge as a Tale

A Discursive Space

Springer

Rafal Maciag
Institute of Information Studies
Jagiellonian University
Krakow, Poland

ISBN 978-3-031-69819-4 ISBN 978-3-031-69820-0 (eBook)
https://doi.org/10.1007/978-3-031-69820-0

This Springer imprint is published by the registered company Springer Nature Switzerland AG
The registered company address is: Gewerbestrasse 11, 6330 Cham, Switzerland

To my children: Suzy, Stan, and Joan.

Introduction

Knowledge is a concept which involves a great effort to explain its meaning. A work that illustrates this effort really well is Mark Burgin's *Theory of Knowledge. Structures and Processes*, which the author of this book considers an illustrious example of both humility and panache. This effort translates into an extensive library of publications addressing the matter of knowledge. In this book, I do not intend to summarize them. Rather, I follow my own narrow path through the dense forest that these works form, to refer to Umberto Eco's excellent and simple metaphor, leading to a predetermined goal—the aforementioned idea of discursive space. I have arranged the chapter titles using a number of select figures acting as their good guardians and as guides at the same time. However, they quickly get lost in the mazes of the corridors they have discovered, which become populated by other characters and run further towards the future. It needs to be added here that in order to cover the expected distance, which I am about to describe, one has to run many special sections dealing with certain specific issues. My hope is that they will not only form some kind of a consistent whole, but also provide a solid dose of knowledge about each other, offered in a concise and accessible form.

The story starts early and runs along a peculiar path at the beginning. This path is called a "digital transformation". This is the name of a seemingly uncomplicated phenomenon, which in fact hides little-known, very important, and still active sources of knowledge, to which the first part of the book is devoted. On the one hand, it defines the fashionable process of evolution of the economy and society in that order, which is the result of the mature implementation of the opportunities and techniques created by digital technologies. Mature, i.e. one in which these technologies create successive levels of applications, creating new phenomena and new areas of their use, such as the Internet of Things (IoT), big data, or artificial intelligence. It should be remembered that these terms are ambiguous, but are most commonly understood as certain economic products. They are also the source of great transformations not only in the economy but also in social life.

On the other hand, a literal analysis of the concept of digital transformation reveals events in the relatively recent past that dramatically affect the process of understanding the nature of the world and the ways in which this process takes place. Its origins are lost in the deep past, but it becomes especially intense in the nineteenth century, especially towards its end, and carries over with the same intensity to the beginning of the twentieth century. It is a story about mathematics and geometry, and about its basic being, which is the number, whose good and uncontroversial description we do not have until today. Mathematics becomes a model for analyzing the world and, at the same time, for its description, which accompanies more and more insightful, subtle and sophisticated ways of doing it, which make it difficult, inconclusive, and ambiguous. There is no paradox in this, but the usual process of digging deeper and deeper into the problem, which does not lead to solutions, but multiplies questions.

However, a very clear cognitive paradigm emerges from this confusion, which is surprising and contrary to common sense, but extremely fruitful. It is based on the idea of an axiomatic system, i.e. a certain innovative and revolutionary descriptive structure that changed the methods of mathematical or geometric description. It allows one to build a description system whose only structural principle of correctness is internal consistency. This means that the principle of representing the world, i.e. confirming this correctness by empirical means, is rejected. In fact, the process of describing is reversed: First a theoretical system of description is created, which can then be applied empirically. This method, inconsistent with common sense and previous practice, whose modern father was Galilee, turns out to be extremely effective. Above all, however, it changes the approach to the nature of cognition, making the theoretical abstract more important and the first step on its way. Inevitably, theoretical abstractions are becoming more and more bold and lead to such unusual proposals of cognitive constructions as a system, network, or complexity.

What is more, a new way of understanding the mutual positioning of theoretical reality based on conceptual abstraction in relation to the reality of practical experiential beings appeared. A new version of the encounter between the world and the subject emerged, clearly and distinctly formulated by Descartes. The new idea abandoned the claim of truth or uniqueness formulated by Plato. Instead of one truth, numerous contextualized interpretations appeared. The twentieth century was a picture of the explosion of various variants of these contexts, which took on a predominantly social character. Thus, a story of knowledge that is not philosophical speculation but is based on the observation of historical and social events. It is a story about knowledge that determines social processes, defining the practice of life and constantly changing. Its nature and the chosen way of description cause this story is the result of choice and interpretation, which is usual for the story as such, as Umberto Eco described. This is also the nature of the second part.

Thus, in the chapter beginning part II, number 3, I describe the key place where the problem of the world's inaccessibility to direct insight is revealed with all its force: language. At the same time, the story of knowledge in this part takes its basic shape, in which important single characters become the axes of its individual stages. Although similar anxieties about language had appeared earlier, it was only Ludwig Wittgenstein who conducted a thorough and dramatic analysis of it in two works that turned out to be one of the most important philosophical visions of the twentieth century: *Tractatus Logico-philosophicus* and *Philosophical Investigations*. The dramatic effect stems from the fact that they generally differ in the way in which they understand the language. The evolution of Wittgenstein's views proceeded from the logical discipline of language to a vision in which it appeared as a specific, complex, and continuous game or labyrinth. In this situation, the relationship between language and the world, which emerged at the end, also became a new epistemological model and rhymed perfectly with constructions such as network or complexity.

The next chapter of Part II, number 4, begins with a mention of Edmund Gettier's achievement, I point to a certain overexploitation of the notion of knowledge, which (rightly) seems to be too simple and, indeed, is underlain with an error that can only be remedied by moving the reflection to a somewhat more general and abstract level. This introduction, somewhat anecdotal in that it recounts a discovery, will show the rapid and latent flow of new thoughts that are about to be revealed, heralded in the field of the humanities by the like of Michel Foucault. But this will happen only in Chap. 7 because in earlier parts, playing with the dynamics of historical time, stretching or shrinking almost relativistically, but respectfully approaching the order of birth of a new worldview, I will talk about knowledge undergoing a dramatic transformation in the way it has been treated and perceived. This process will run through the area of a broader line of thought concerning man and man's self-organization in the world, and thus through the paths of sociology and its branches that describe humans' efforts aimed to enable them to comprehend the surrounding world.

It is no exaggeration to call this process the emergence or creation of a new human world, and the wave of new ideas and events that engulfs and subverts it, leading to new arrangements in the 1960s and 1970s, is very powerful—and its impact continues to this day. This is more or less what I discuss in Chaps. 5 and 6, delineating thus my own symbolic heroes: Ludwik Fleck, who prophetically reminded science of its social place, at the same time destroying the metaphysical (epistemological) hubris in which it had been wallowing. For what else could a conviction of exclusive access to truth be? A conviction similar to a religious one, only even more egotistical. Masuda acts as a symbolic exponent of this breakthrough, making mankind realize that it has finally fulfilled a prophecy well known in almost all cultures—meaning that it has have set in motion forces that they are unable to control and that will ultimately destroy them, which can mean a profound transformation of the entire civilization across all of its levels and areas. We are, of course, talking about digital technology here.

At this time, knowledge is thrown on the table, becoming the object of curious inspection and study as a potential raw material—into which it eagerly soon turns. Its old and unusual nature is uncovered by Foucault, revealing an interior that is fragile and connected to the human environment in an extremely complex way, as described in Chap. 7. This is the very body of knowledge picked up by Lyotard, who plays the symbolic role of a visionary here—he sees the great causative power of knowledge in the ethereal substance of it, already materializing but still alien. It is no longer a piece of strange matter lying motionless, but a tissue torn from its social body, acting simultaneously as its bloodstream, nervous system, and skeleton—and actually forming the entire body thereof as well. This is the conclusion to which the final—eleventh—chapter offers. But for the time being, in Chap. 8, as groundbreaking as the previous one, we celebrate the act of making use of the power source inherent in the substance of knowledge.

The next two Chapters, 9 and 10, follow a bit different pattern but still manage to retain the adopted convention. They are based on a deep belief that the metamorphosis of knowledge has already taken place before our very eyes and that it is necessary to highlight those sources of thought that will make it possible to describe it in its new form. There are two of these sources. And even though they differ in terms of their subject matter, they are similar in the way of reasoning—abstract, advanced, piercing through and fleeing the hard material surface of the world to take advantage of constructs built in other spaces—far more general and basic. They are discourse and space. They are also difficult, not obvious, and incompatible with common experience, so they require separate research because they simultaneously open up completely new possibilities for understanding and describing the world.

In the structure of this line of thought, the two chapters provide direct methodological preparation for the construct that appears in the final chapter, 11. This already mentioned chapter presents the idea of the so-called discursive space, which has been conceived as a descriptive model of knowledge that uses the phase space construct, combining qualitative and quantitative approaches. This idea has already been presented in several articles, but in this book, it is complemented by an extensive introduction and rationale not published elsewhere. The chapter in question is the culmination of the book and supports the concepts and structures described in previous chapters. The author's intention is that this final chapter should arise from the preceding ones, gathering the necessary content along the way like a river from its tributaries, in order to eventually turn a single stream, shaped by circumstances, yet running in an orderly, arranged manner. This is a risky structural assumption because such an order does not refer to the familiar composition of reasoning: assumption—hypothesis—proof. Thus, it does not seek to answer the question of how the world really is and validate that answer as relevant. It merely proposes a cohesive and internally non-contradictory proposition, which is, however, quite free within that cohesiveness and multiplies possibilities rather than closing them off and excluding them in the search for the right ones. It is a tribute to the unbridled nature of thought and necessarily a tribute to its opus—knowledge.

Finally, I would like to thank the important people who have provided me with the necessary and valuable inspiration: Marek Burgin, Marcin Schroeder, Wolfgang Hofkirchner, Piotr Bołtuć, and Marek Hetmański.

Kraków
11.02.2023

Introduction No 2

Now that we know what a book consists of, it is time to introduce a different type of introduction. I will try to explain the shape of the book: a digressive and multi-threaded story. Such is the case in this introduction, which is not a uniform reasoning, bound by an iron discipline of conclusions. It is a growing set of different circumstances that at some point will exceed the critical mass and become a coherent wholeness. Through this wholeness, the sun of final meaning of the notion of knowledge should finally light up.

It is not the phenomenon of knowledge itself that is recalled here, but the notion that names it. This is the basic analytical decision that determines the most general research perspective. This notion is like a footbridge that connects the busy and deceptive world of language with the world itself. Provides tools to talk about our environment, understand it, record it, and ultimately include it in our own world at all levels of its generality. The unusual features of language have been observed since antiquity, but only since the beginning of the twentieth century has it become the subject of advanced reflection, the canonical examples of which are the works of Ludwik Wittgenstein and Ferdinand de Saussure.

Language is, of course, a function of the social nature of human relationships. Everything that happens in language has its source there and ultimately shapes these relationships. This mutual and complex process can be understood as the basic and even the only way of generating all experience and at the same time shaping human societies. This is how it was perceived, among others, by Michel Foucault, a researcher who made this fact the core of his extensive philosophical reflection. His approach is the main inspiration here. Notions appear at the forefront of this process because they go to impact as the first surface of the world they try to capture and—as imperfect, momentary, and local—they are always doomed to failure. They perform their duties so poorly that they remain inaccurate, partial, and context-sensitive. They are also subject to the process of change and reconstruction.

This fact became the basis for concern about the phenomenon of knowledge and the axis of thinking about it. The reason for this mode of approach to knowledge is obvious. This book, its introduction, and all other extremely rich analyses and findings on knowledge cannot avoid the mediation of language. Talking about knowledge is always using a certain notion that refers to knowledge as a phenomenon. There is a vast literature dealing with the phenomenon of knowledge, beginning with Plato's dialogue Theaetetus, which contains its famous definition, also discussed in the book.

Language, which is an environment of notions, has roots as deep as the existence of the people and societies of which it is the fruit and companion. It is a process and an event, in which it is similar to the notions in which it consists. The notion of knowledge is, of course, of the same nature. We are talking about words, the most basic tissue of our speech. When we observe how they appear, how they operate in each act of using them, and how they change, we can naturally ask two questions. The first question arises about the process of historical change in the meaning of a notion and the relationship that connects it with its counterpart in the world. This question immediately turns into a task of examining the historical circumstances brought about by the passing years during which this change takes place.

The second question naturally concerns the current state in which the notion of knowledge appears. This task is paradoxical because it attempts to freeze the changing shape of the notion in a stable image. However, the reasoning presented in the book tries to avoid this mistake. For this purpose, it proposes a model that takes the analysis of the concept of knowledge beyond this dialectic using a descriptive structure, which is a dynamical space. This model is called discursive space and is described at the end of the book. Although the construction of dynamical space comes from the field of physics, it is also well represented in the social sciences. However, dynamical space is based on the idea of describing variable and moving processes. Therefore, it must be supplemented with a justification and description of the dynamics of the notions it deals with.

The first question and task at the same time, the one that respects the variability of the concept of knowledge, therefore remain the axis of the research work. The reasoning carried out in the book is inspired by the idea of complexity, the dynamics of which is based on incalculable, mutual interactions, going far beyond the usual cause-and-effect order. Rather, it is based on persistent tracking of these dense contexts in which the notion of knowledge is entangled, revealing itself thanks to them, in confrontation with them, in mutual dialogue. A careful reader will notice an attempt to apply also the best analytical traditions of Hans Georg Gadamer and Michel Foucault.

If we adopt this mode of reasoning, two important circumstances immediately appear, which we will discuss in turn. The first circumstance is that an extraordinary, yet obvious, transformation of the object of our investigation is taking place before our eyes. The second one concerns the rapidly expanding perspective from which we look at language and its notions, but more on that in a moment. Let us start with the mentioned transformation. It occurs because the more we entangle the problem of knowledge in the contexts in which its notion appears, the more the phenomenon of knowledge diversifies and enriches. This is the result of an extraordinary feature of language, according to which notions gain meaning only when they participate in the game of their direct use. This is one of Wittgenstein's most famous observations—the language game that creates the meaning of concepts is a whole consisting of language and the activities into which it is woven. It is obvious that the more often these notions are used and the more numerous their contexts appear, the richer and more complex their meanings become.

It seems at first that starting the search for the answer to the question of what knowledge is from a notion instead of a phenomenon is a false inversion of the usual order of experience. This order is based on a natural progression from the wealth of phenomena revealed to the observer to subsequent attempts to organize and explain them. However, here the order is reversed: it is the reconstruction of the notion, its participation in [new] activities in which it can be used, creates a new way of understanding it, which leads to organizing the observed world in a new way. Therefore, language comes first, not experience. Therefore, the first is the descriptive system, not the object of description.

This seemingly paradoxical approach, in a slightly humorous and simplified way, presents one of the most important cognitive inventions, which was the axiomatic system that appeared in the area of geometry and algebra. I devote a lot of space in the book to describe its uniqueness, which consists in opening completely new and dazzling ways of looking at the world. He abolished all limitations of imagination, which for centuries had been forced to keep thought consistent with experience. The revolution took place at the turn of the nineteenth and twentieth centuries, completely changing the approach to mathematical and geometric entities, quickly becoming an inspiration for the humanities.

The freedom to create new worlds, which was also realized in art as pure abstraction, allowed for the formulation of one's own rules for generating theorems in geometry and algebra. The only requirement was to maintain internal coherence, defined at the beginning. It is therefore not surprising that formalism, which meant the complete abandonment of the representing the world in cognitive constructions, became of great importance in the process of creating theories, both in mathematics and art. Their justification was entirely contained in the descriptive system itself, without the need to look for evidence in the world outside.

The lesson of this freedom, which also meant the need to completely reconstruct the idea of truth, was also a lesson of the existence of many parallel possibilities of understanding and describing the world. The number of possible geometries and mathematics is virtually endless. This conclusion was both creative and bitter. It also necessarily raised a fundamental systemic doubt in the descriptive systems that had been operated so far. Their claim to exclusivity, based on consistency with experience, turned out to be illegitimate. This doubt has generated new, paradigmatically deep research approaches. One of the fields in which it was realized was, of course, language, led by the vision presented by Wittgenstein.

Wittgenstein's concept should be seen as the ultimate reconciliation of both sides of the cognitive situation: language and the world. The language game consists of both of these elements, and yet there is no prior clear recipe for the nature and structure of their combination. Wittgenstein initially assumed that language was a coherent formal system that followed the rules of logic, reflecting the similar nature of reality. Later he completely denied this belief. The place of logic and coherence has been replaced by play and affinities that appear locally, so there is no point in searching for their metaphysical foundations. In such a situation, the study of meanings becomes collecting the implementation of notions, searching for momentary regularities that constitute families of concepts, variable, and blurred. This means many simultaneously existing affinities, or, in other words: systems of justifying them.

I have devoted a lot of space in the book to illustrate the description I have just given of how language works. One of the most important paths of reasoning leading to it led through the concepts of Michel Foucault and Jean François Lyotard. Both of them were convinced that there must be some structures that exist beyond notions and their systems. These structures produce meanings in various ways in parallel, replacing themselves or co-existing, and changing during the historical development. Foucault used various terms, of which the most famous and the most widely developed is discourse. Lyotard introduced another term: narrative. In each case, the aim was to identify overarching and at the same time local meaning systems that combined consistent instances of lower levels of language, such as notions or sentences, into internally coherent meaning constructions. They both also had no doubt that these systems were also coherent articulations of certain knowledges. The plural form of this notion is used deliberately to ostentatiously break away from the otherwise suggested fiction of one knowledge, true by definition.

Both discourses and narratives provide sufficient justification for certain knowledges that exist simultaneously in a numerous, practically infinite set of variants. These are descriptive systems similar, even analogous, to the formal axiomatic systems that destroyed the hegemony of Euclidean geometry and traditional algebra. Confirmation of this thesis can, for example, be found directly in Lyotard, but the similarity of the dynamics of thought struggling with the curse of the compliance with the world, thought that rebels and liberates itself, is striking. You can see it sailing towards unknown lands, boldly and without fear. All these poetic metaphors are justified because the challenge that appears before the conscious mind is great. It makes it an adversary equal to the world, although it ultimately leads to his defeat because formal systems, as Karl Gödel proved, are ultimately, in their fullness, unattainable.

Language, however, is not a formal system; it is its living negation, running through a multitude of various configurations of meanings and various logics immediately, in one act of its use. This is why poetry exists. And it leaves traces of its appearances and presence in the universal memory of its creations such as discourse. This memory is, of course, not eternal but multiplies associations and contexts. This is why Wittgenstein compared language to an old city with an ever-expanding tangle of streets, allowing one to wander to one's destination in an ever new and different way. In each such journey, new sights and encounters appear, making it an exploration of the unknown, that is, what has not been part of the journey before. New curves and straights appear. The course of the journey changes, forcing its reconstruction. The notion of knowledge is subject to such dynamics as that I discover in the book showing how different and new roles it can play.

This complex processing of meaning would not be possible without the property of language with which we began this introduction. This property was considered the second key factor in its existence and operation. So far we have focused on the internal dynamics of notions that animate it, but now we need to change the perspective and look at language as an extremely extensive event existing in the social environment. Language is created between people and remains only between them—this is obvious, and they, thanks to language, among others, establish relations of a very complex nature. Language is an event of unprecedented, massive scale. Bruno Latour understands the notion of social as a movement, a displacement, a transformation, not a real and stable situation or object. Language inherits this dynamic, which is undoubtedly complex in nature. It means that, among other features, language does not allow for tracking and accurate reconstruction of mutual dependencies expressed in simple cause-and-effect connections. There are too many of them. And they develop too many relationships. So much that by definition they escape the analytical apparatus, allowing themselves to be observed only as certain wholes heading in a certain direction. In this practical way, a working environment for notions is created.

Foucault, Lyotard, and others were perfectly aware of the described mechanism of language operation, which could be described as "social", although they did not call it complex in the sense that this concept was acquired only in the 1970s. They also had no doubt that it is a vehicle of knowledge that is revealed and shaped in all expressions of language. Researching the notion of knowledge is completely justified from this point of view, although less expected due to the role and perception of this phenomenon. It was enough to draw conclusions from the collected premises and wander with the notion of knowledge through its articulations, which turned out to be an extensive and branching journey, and then write down the experiences of this journey.

In this wandering, expectations and decisions (and therefore the order of wandering) are constantly reconciled with unexpected encounters and discoveries. It is therefore difficult to maintain complete discipline in the argument, although, of course, such effort accompanies the journey (directions, documents, maps). On the other hand, you cannot avoid the poetics of the story, which consists of adventures, surprises, and sometimes pulls the writer towards alleys, starting new, complicated threads. At the same time, the story remains an act in which the listener or reader is indispensable. Although it is intended to be monologue-like in nature, the story needs the presence of another person, and thus becomes a strong element of what is social. Finally, the story, freeing itself from the nature of proof in reasoning, on the other hand opens itself to uncertain knowledge that is difficult to gather and organize. However, it opens the possibility to speak openly and freely, even make things up, but not lie, demanding only tender acceptance of weaknesses. In this situation, it remained only to tell a story.

Contents

Part I
Digital Transformation

Chapter 1
The Digitality of Digital Transformation

The notional cluster reading 'digital transformation' became a popular search in Google only in 2015.[1] Before that year, its search trend line appears to be flat, adjoining the horizontal border of the chart. Of course, it is only a curve illustrating the relative level of interest compared to the maximum level, not a chart based on absolute numbers. Plus, it only covers the period from 2004 onwards. But its shape is distinctive; it climbs consistently from the said year 2015, and the maximum level comes at the very end, in 2019, so it is quite likely that the rising trend will continue. The chart shows two qualities clearly: the presence of a significant trend and the relatively close initiation of the period of an increased level of interest. The notional structure of digital transformation is therefore much younger than the technical innovations it is associated with, and has been in the broader intellectual circulation, so to speak, for only a few years now. This means that the need to address some fragment of reality using a new idea has appeared relatively recently as a cognitive action, resulting most likely from a growing impression that the old, established processes entered a completely new stage at some point—a stage calling for a new naming tool, a tool enabling the inclusion of new ideas into the existing body of knowledge. This new notion therefore tells us a lot about the ability to understand the world, and not about the world itself.

"Digital transformation" is a cluster of two words. According to the Oxford University Press dictionary (Lexico.com 2020), the first word—"*transformation*"—comes from "(to) *transform*", which originates from the Latin verb "transformare". The latter is a product of two words: the prefix "trans" and the root word "form", which may function as both a noun and a verb. The original Latin word "trans" means "across". The word "form" comes from the Latin "formare"—"(to) form", which in turn stems from the Latin "forma". The latter, however, does not close the spectrum of meanings but actually opens it. This spectrum of meanings starts with

[1] https://trends.google.pl/trends/explore?date=all&q=digital%20transformation (accessed on: 8/15/2019).

R. Maciag, *Knowledge as a Tale*, https://doi.org/10.1007/978-3-031-69820-0_1

a deep reflection on form as discussed by Aristotle in his "Metaphysics", where the philosopher argues as follows: "by 'form' I mean the essence of each thing, and its primary substance" (Aristotle 1933).

The idea of form, as it appears, is deeply ingrained in metaphysics: it defines the essence of things, meaning their manifestation in the world, which constitutes everything they are, the roles they perform, the relationships they form, etc. This essence shapes their matter, which is the second fundamental aspect of the Aristotelean ontology next to form. Moreover, form is rooted in the primary and only cause of everything, so it does offer some transcendental sanction set beyond the boundaries of the human world. Changing an object's form is therefore a dramatic interference with this object, an alteration of its essence, an act of placing this object among other objects and arranging all these objects in some specific order. Transformation can then be understood in this context as an act of crossing the existing essence of a given being. An act of re-forming its presence in the world and in relation to the world. If we do not pinpoint such a being, it could seem it concerns everything, all, a fundamental and limitless traverse. One needs to think very carefully if perhaps the said intuition does not underlie the word in question when this word appears in the cluster "digital transformation".

Moreover, transformation is a certain deep-rooted and constant property of the world—which stems from its fundamental and imperfective nature. It describes something that lasts, something that is both a course and a process by its very essence. All in all then, when considered as a notion and a vision of the world at the same time, it is a paradox. In a single description of the world in the form of a word, it encompasses a certain continuity, meaning it makes this word divisible and discrete, thus interfering with its essence. The paradox is not necessarily a mistake, but rather a guideline offered to us by our system of reasoning, suggesting that we have encountered a problem the system is not able to handle and thus requires switching to a different, more general level of thinking (Sorensen 2005). In the case of transformation, the paradox is about a certain disproportion between the world and the language used to describe it and the technique utilized by the language to deal with the problem. It is clear that the said disproportion—or the inaccuracy or incompatibility of the language we use—is the main and frequent flaw of this language. A reflection on the mechanism underlying these properties of language will make it possible to determine its ability to describe the world. Our ability to describe the world. The problem in question, which concerns fundamental matters after all, will be raised in the chapter addressing the general bases and sources of reasoning.

The second word in the phrase "digital transformation": "digital", acts as a *differentia*, meaning that it offers a distinction which makes it possible to define some specific part of a greater, superior whole (genus)—being transformation. This order arises from the classical idea of definition, which is based on the theory proposed by Aristotle in "*Topics*" (Bayer 1998, 487). The type of relationship included therein shows that the idea of transformation is more general, ranked higher in the hierarchy of the description of the world. This is otherwise clear in light of the sense stemming from the Latin etymology of the word "*form*". It describes, after all, certain fundamental features and qualities of the world, with other features and qualities being additional and secondary to the former. This is a very important conclusion. One that stresses the part of the phrase which is usually somewhat disregarded. Transformation as an idea and a phenomenon is obviously older, more fundamental than its digital face, so to speak.

The word "digital" offers similarly interesting and rich coincidences. It comes from the word "*digit*", originating from the Latin "*digitus*"—"finger", but Merriam-Webster adds that it might be *perhaps akin to the Greek* "*deíknȳmi*", "*deiknýnai*" (Merriam-Webster.com 2019b), and the word that means "(to) *show*", "*(to) point out*" comes with a broad spectrum of continuations in many languages (Merriam-Webster.com 2019a). Varanini provides an in-depth analysis of connotations related to these types of meanings, which form an interesting network of related meanings. Showing, pointing, associated with the finger, has its source in the Proto-Indo-European language, and can be found in Sanskrit as well. For Varanini, it is a sign of a fundamental significance of the act of showing, which makes it possible to differentiate the elements of the world from each other. Its presence in the context of numbers is therefore certainly not coincidental, especially since the English word "finger" has its roots in Proto-Germanic, to which it was imported from the Proto-Indo-European word "pénkwe" ("five"). Showing and counting seem to be closely related acts. The former precedes the latter in the sequence of organizing and naming the surrounding world (Varanini 2018, 15).

The word "digital", apart from its anatomical sense referring to fingers, not only points to digits per se but also focuses on those which form the binary system, meaning only two of them: 0 and 1 (Lexico.com 2022). Today, what can be defined as belonging to the digital domain, is binary by default. The binary system—but also the very principle of the establishment of such a system—is the foundation of modern computing. Two states can reflect many different values—not only digits but also the dichotomy between true and false. They can be shown as an opposition of presence and absence. Software engineers do not need their ontological interpretation, which certainly underlies the said opposition. All they need is the presence or absence of electric potential, which shapes physically this dichotomy and enables manipulating its symbolic senses: digits and logical states of true and false.

The Merriam-Webster dictionary highlights another—clearly distinct—semantic context related to the word in question. One which continues the emerging coincidence with the computer reality. The context is underlined clearly by Floyd in his textbook on digital electronics: "**An analog quantity** is one having continuous values. **A digital** QUANTITY is one having a discrete set of values" (Floyd 2014, 16). This discreteness, meaning a division into clearly separate parts forming a stream, mimicking an uninterrupted flow of a real, source phenomenon, is the technical quality on which all technical digital solutions are based—contrary to the analog ones. If we take a closer look at the case in question, the logic of quantification is an immanent trait of a number and a logical state. It can be challenged, assuming an infinite range of a number as a certain abstract and blurring the logical border between true and false. This is how the dialectics of continuum and singular states, an issue of considerable importance at a certain level of advancement of mathematics, comes into play. The world seems to be continuous, but counting in in its different aspects appears to necessitate divisions. An example is the simplest physical processes, such as the movement of the body in space. Describing these processes in mathematical terms started from Newton and Leibniz, who introduced a very basic mathematical tool: the infinitesimal calculus (Kline 1972, 342). Ancient Greeks were not familiar with the contemporary notion of continuity. This lack of knowledge was what Zeno of Elea's paradoxes were based on. Continuity has one more flaw: it disappears in infinity, which is a certain indivisible whole. This way of existence was first understood only in the nineteenth century thanks to the concept of Georg Cantor (Kline 1990, 992). And thus we are moving inevitably closer to the problem of number.

Number and Mathematics

The foreground of the word "digital" is occupied by a being called 'digit' which carries a tale of a great intellectual adventure of man. Georges Ifrah's book entitled "*Histoire universelle des chiffres*" (Ifrah 1981) is a monumental work that tells this tale with great zeal. Its author has no doubt that it is story of mankind, dating back to its very beginnings, connected with the very basic ability and need to count, present in all cultures and with its roots hardly traceable. The act of counting in the times when it emerged was based on the use of various small objects, sometimes made for this purpose, such as rocks or clay figures. Its progressing development, the growing range of its applications, and the increasing level of its complexity led with time to the invention of a symbolic representation of counting tools in the form of digits. Irfah claims that the oldest known symbolic pictograms representing digits were created by the Sumerians in Lower Mesopotamia about 3200 BC (Ifrah 2000, xx), which generally coincides with the appearance of writing in this place and at this time.

Counting, numbers, numerical systems are all a great record of the intellectual achievements of man, being a sum of inventions created in different parts of the world, whose evolution led to the emergence and development of a discipline of extraordinary importance—mathematics. Mathematics offers an unignorable context where the presence of a symbol such as a digit takes a relatively most advanced form, reflecting a certain state of understanding of the world that accompanies it: a certain knowledge, a view of convictions or beliefs. Wilder argues as follows: "mathematics is not something which is by its nature universal, absolute, or foreordained; it is subject to laws of development and influence from other cultural elements much as are arts and sciences in general" (Wilder 1967, 285). It is hard to imagine digital transformation without computers, and computers without mathematics. The dash accompanying the gradual progression in this branch of knowledge is characterized by a dynamics of its own.

The brief history of mathematics, as told by Ifrah (2001, 72), begins around the eighteenth century BC in Mesopotamia, and then shifts to ancient Egypt to emerge in Greece in the fifth century BC, where it develops thanks to Pythagoras,

Archimedes, and Euclid. At the beginning of our era, the development of mathematics is joined by China followed soon after by India which contributed the notion of zero—the opposite of infinity; an idea based on mystical depth, a concept unknown in Europe at the time. In the ninth century, the process of the development of mathematics was joined by Muslims: Al-Khwārizmī (the Latinized version of his last name gave birth the word "algorithm"), Abū Kāmil, Al-Karajī, and others. In 1202, Italian mathematician Leonardo of Pisa, also known as Fibonacci, wrote his "*Liber abaci*", which documented the achievements of Arab and Greek mathematics, becoming the foundation of European mathematics for the next three centuries. The development of mathematics moves to Europe, and Western civilization gradually overshadows the rest of the world in the achievements in the discipline because—as it is easy to see—political history is reported as part of the history of mathematics with absolute precision, repeating the image of politically and civilizationally dominant parts of the globe.

The development of mathematics in Europe in the nineteenth century and at the beginning of the twentieth century brought surely the most advanced and abstract way of perceiving numbers, which is connected with the most fundamental questions regarding mathematics and its beings. The nineteenth century is also a time of revolutionary changes in mathematical ideas. The earliest geometry is subject to a dramatic transformation manifested in full bloom in the idea proposed by Hilbert in 1899 in his *Grundlagen der Geometrie* (*The Foundations of Geometry*). The idea of axioms was known since the age of Euclid, but Hilbert defined them in a clear way as a self-aware, arbitrary, own, and free selection of assumptions on which a further formal reasoning is to be based. The idea was put forward even some time earlier by Peano. In the introduction to his work, Hilbert states that his investigation "is a new attempt to choose for geometry a simple and complete set of independent axioms and to deduce from these the most important geometrical theorems" (Hilbert 1950, 1). Peckhouse makes the following remark: "the basic objects of his [Hilbert's] system of axioms are "thought-things", i.e., products of human thought (…). Geometry now becomes a speculative discipline, its relation to intuition becomes irrelevant, or, as Hans Freudenthal took it, the connection between the reality and geometry is cut" (Peckhaus 2003, 142).

Hilbert explains his approach in a letter to Frege in the following way: "every theory is only a scaffolding or schema of concepts together with their necessary relations to one another, and that the basic elements can be thought of in any way one likes. If in speaking of my points I think of some system of things, e.g., the system: love, law, chimney-sweep (…), and then assume all my axioms as relations between these things, then my propositions, e.g., Pythagoras' theorem, are also valid for these things. In other words: any theory can always be applied to infinitely many systems of basic elements" (Frege 1980, 40).

The step taken by Hilbert eventually shaped the manner of reasoning grounding its logic only on internal relationships, inferences. Such reasoning does not need any external references, and especially no justification based on experiences or even any relationship with the world (reality). It is based solely on the internal order adopted on the basis of the structure of axioms. Humphreys refers to this kind of technique as syntactic axiomatization (Humphreys 2004, 552). Murawski writes that "in this way geometry become pure mathematical theory. Axioms were not treated any longer as evident and necessary statements. The question about their truth lost its meaning and sense. As axioms any sentences could be adopted. The main problem was now not the problem of consistency of given axioms. Geometrical deductive systems became uninterpreted axiomatic systems various interpretations of which are possible. In this way, the traditional philosophical view which regarded geometrical knowledge as synthetic a priori knowledge of our world has been decisively refuted" (Murawski 2004, 582).

Also, the works of Richard Dedekind, Gottlob Frege, and Giuseppe Peano appearing in the second half of the nineteenth century and addressing numbers "did lead to a reshaping of mathematics if not the whole world-view" (Hodgkin 2005, 215). They brought about a "crisis of foundations" regarding the sense and the justification of essential mathematical beings. They lost their nature of intrinsic, self-contained beings that could be described, defined, or used: "the objects of mathematics were not actual things-in-themselves (as one thinks of a triangle, say, or the number "7"), but the rules which they obeyed" (Hodgkin 2005, 216). This meant a departure of mathematics from the world of experiments, and an immersion in a reality of almost any kind of abstract constructs. Many years later, Morris Kline referred to the situation as "the loss of truth": "By 1900 mathematics had broken away from reality; it had clearly and irretrievably lost its claim to the truth about nature, and had become the pursuit of necessary consequences of arbitrary axioms about meaningless things" (Kline 1990, 1035). Mathematics, viewed thus far as a stable structure, one making the world stable as well, changed into an unanchored balloon: untamed, unpredictable, and dangerous.

This, of course, led to a crisis that triggered a search for solutions and a need for fundamental questions and answers to problems concerning mathematics itself. Such circumstances give birth to the idea of a purely conceptual machine, acting as an important element of reasoning—not an actual, real object. An idea that will lie at the heart of a computer later on. The machine was invented by Alan Mathison Turing, named after his name and described in his famous 1936 article entitled *On Computable Numbers, with an Application to the Entscheidungsproblem* (Turing 1937). The machine is used to illustrate the dynamics of actions expressed as subsequent steps and taken in the form of manipulation of arbitrary symbols. These are simple steps as they actually involve only being written down and erased in some specific order. The steps are materialized as a mechanical procedure that moves forward by means of pre-assumed simple conditional rules: identify the symbol and erase it, change it, or leave it, and then move to the next symbol. The machine, simple yet highly ingenious and offering an incredible potential manifested in the interpretation of the symbols it utilizes, is a basic subject covered in all IT textbooks—including *Algorithmics. The Spirit of Computing*, a classic by David Harel (1987), reissued many times, where he wrote as follows: "Turing machines are capable of solving any effectively solvable algorithmic problem! Put differently, any algorithmic problem for which we can find an algorithm that can be programmed in some programming language, any language, running on some computer, any computer, even one that has not been built yet but can be built, and even one that will require unbounded amounts of time and memory space for ever-larger inputs, is also solvable by a Turing machine" (Harel and Feldman 2004, 228).

Turing's concept was an answer to the fundamental mathematical problem set by Hilbert mentioned earlier. During his lecture given at the Second International Congress of Mathematicians which took place in 1900 in Paris, Hilbert presented his famous program in the form of twenty three unsolved mathematical problems. The program was "an attempt to justify the classical (infinite) mathematics and to save its integrity by showing that it is secure" (Murawski 2010, 29). Implementing the program was to stabilize and justify mathematical procedures and constructs, and to be more precise—build the mathematical foundation upon a set of certain specific axioms and rules of inference. Problem 10, described in more detail by Hilbert 1928, came down to a question Penrose summarized as follows: "is there some general mechanical procedure which could, in principle, solve all the problems of mathematics (belonging to some suitably well-defined class) one after the other?" (Penrose 1999, 34).

Hilbert's program failed eventually when, in 1931, Gödel proved that it was impossible to establish such a system (Nagel and Newman 1958). Turing gave a negative answer to Hilbert's question asked under problem 10 as well. Penrose explained his line of reasoning in great detail, using a language that was far from mathematical. He summarized Turing's conclusion in the following words: "no algorithm exists for deciding the question of the stopping of Turing Machines" (Penrose 1999, 63). Similar conclusions were arrived at by Alonzo Church at a similar time, although he based his inference on a different line of thought.

When describing a Turing machine, Penrose resorted immediately to binary system as the relatively simplest symbolic system, which makes the machine a calculation tool at the same time. Modern computers are all based on a binary system. This is very convenient because the numbers can symbolize logical states (true and false), and numbers such as 1 and 0 become interpreted in logical terms. The application of logical calculus proposed by George Boole (1854; Davis 2000, 32) makes it possible to perform calculations on logical values. Moreover, in practice the existing boundaries become blurred. This quality is the basis of editing of a digital image, which is represented in the form of a stream of numbers encoded in a binary system. The same numbers can be used to perform logical operations such as obtaining a product or a logical sum. This is how many filters enabling the editing of digital images work. In the physical reality of a computer's operation, numbers and logical states take on the same form of manifestation—the differences in electric potentials.

But a mathematical binary system shall not be understood as merely a technical facilitation. A decision to utilize it involves more serious consequences. Consequences regarding the view of the world imposed by a given decision—a world fundamentally and transcendentally tied with numbers. Issues related to such type of context have accompanied binary system since its beginnings, appearing directly in the descriptive structures available throughout time. Ifrah covers the complex circumstances of the emergence of this system at length. The system is said to have been created by Leibniz, who saw it in the Chinese system of the *yin* and *yang* symbols. He could become familiar with it thanks to *I Ching*, a book he received from his friend who was a mathematician and a missionary, and brought it from China after being sent there by Louis XIV. But Ifrah claims that Leibniz was wrong about the allegedly original mathematical value of the symbols he saw. He was inspired, however, to devise the binary system which he described in his work published in 1679, entitled *De progressio dyadica* (Ifrah 2001, 91) (Ifrah 2001, p. 91). The foundation of the concept was a deeper, metaphysical, and religious idea of creation of the world, based on the numerical symbolism of two digits: 0 and 1. For Leibniz, number one was a symbol of both a number and God at the same time, while zero also meant nothingness. The space between nothingness and God—like the range between zero and one—was the area of everything. The universal source of this everything was One, which had both a religious and mathematical meaning. In his 1697 letter to Duke Rudolf August, Leibniz included the following famous sentence: "Omnibus ex nihilo ducendis SUFFICIT UNUM" (Leibniz 1740, 95) (Leibniz 1740, p. 95). The sentence may be translated in two alternative ways: "in order to produce everything out of nothing, one [thing] is sufficient" and "in order to produce every [number] from 0, [the number] 1 is sufficient" (Lenzen 1991, 194). This lexical ambiguity, typical of the baroque at the end of which it came into existence, was at the same time a testimony of Leibniz's deep conviction as to the fundamental, transcendental role of numbers and mathematics in the world, which he described as a religious—and even mystical—experience.

The discussed story about numbers and about the way Leibniz approached them is not about a summary of the religious views of the author, nor about their innovative nature,[2] but about a conviction that when addressing the issue in question, we touch upon the greatest mysteries of the world's existence. This conviction, age-old, has been founded largely on the most basic observations of astronomical cycles, which people have tried to express in strict, clear terms, discovering the mathematical order of the cycles of life and the world at the same time. This fact has made the act of counting reach a transcendental status, one reaching beyond human competence, since the earliest times and in different cultures (Ifrah 2000). Interestingly enough, the said conviction has not been satisfied by the intellectual efforts of subsequent generations, although it has become increasingly difficult as an object of reflection, hiding itself deeper and deeper—also when it left the domain of religion and settled in the field of mathematics once and for all. It seems that out of necessity and given the ever uncertain status of number and mathematics, there must be similar concerns—but expressed by means of a different language—lying at the root of such constructs as digital transformation.

A rather dramatic and not very distant period of modern views on the matter in question taking shape was the said second half of the nineteenth century. It can be described as a period of the dramatic reconstruction of humans' cognitive

[2] A similar observation was made over 150 years earlier, in 1623, by Galileo in his text entitled *Il Saggiatore*, where he used a metaphor comparing the universe to a voluminous book written in the language of mathematics, with letters being triangles, circles, and other geometric figures (Maciąg 2012, 175).

capabilities and knowledge in the field of science, aspiring towards the definition of truth, and within this knowledge—to the definition of mathematics. Those events trigger a development in the domain of arithmetic applied for the purpose of mathematics, meaning a sophisticated and complex process of counting, dating back to around 1870 (Gillies 1982). Out of necessity, it had to be centered around attempts to define number, which involved describing natural numbers as well as numbers from outside this set and posing a great challenge in terms of their interpretation since the age of Pythagoras—like irrational numbers. The deliberations led to deeper and more fundamental questions regarding mathematics alone, which caused the crisis referred to when discussing Hilbert's program. The program was an attempt to reorganize and develop the mathematical thought in general. As Olszewski writes (Olszewski 2009), at a similar period, i.e. near the end of the nineteenth century, the most impactful philosophical concepts regarding mathematics were still in the budding stage. The internal rearrangement of this field of knowledge, given a complete lack of similar attempts made earlier, proceeded almost simultaneously with the consideration concerning its theoretical status and justification. In his monumental work on the history of mathematical thought, Kline argues that "One of the most surprising facts in the history of mathematics is that the logical foundation of the real number system was not erected until the late nineteenth century" (Kline 1990, 979), which meant that mathematicians relied only on an intuitive understanding of real numbers which—after all—encompassed all numbers we use on an everyday basis, meaning natural numbers, integral numbers, rational numbers, and irrational numbers like the number π. The lack of logical grounds meant that there was no formal definition of these numbers, whose existence seemed to be clearly obvious and not requiring any in-depth description.

The most important works providing the foundation for the said programs, considering the problem of numbers as fundamental beings, included: *Die Grundlagen der Arithmetik* (The Foundations of Arithmetic) by Gottlob Frege (1884), *Was sind und was sollen die Zahlen?* (What Are Numbers and What Should They Be?) by Richard Dedekind (1888), and *Arithmetices principia, nova methodo exposita* (The Principles of Arithmetic, Presented by a New Method) by Giuseppe Peano (1889) (Olszewski 2009, 22). They created a certain unintended evolution of the idea of number. Frege believed that "the content of a statement of number is an assertion about a concept" (Gillies 1982, 45), looking for the source of justification of a number—including a natural number as the most primary, fundamental number—in some external reality of logical beings, such as notions and concepts. As Gillies reports, Dedekind did not aim at arriving at a direct definition of a natural number, but rather channeled his efforts "to produce what might be considered as a set-theoretic generalisation of the sequence of natural numbers" (Gillies 1982, 59). He did it by proposing a concept of an infinite set of natural numbers, based on which it was possible to produce a natural number as "a free creation of a human mind" (Gillies 1982, 61). In the case of Dedekind, numbers were also established by means of a reference to some external instance, which he considered the human mind as the source of thoughts. This was the reason he was attributed with naive psychologism.

The most advanced—because the most abstract—and, looking back, the most influential manner of 'consolidating' the nature of natural numbers was proposed by Giuseppe Peano. At the same time, he adopted a certain type of a fundamental approach which served as the basis for said Hilbert and others, which led to a reconstruction of mathematics at the beginning of the twentieth century. Peano rejected any and all external instances such as logic or the mind, which required recognizing their separate and justified nature anyway, making the further reasoning dependent on these assumptions. He regarded a natural number as "an undefined notion which is characterized axiomatically" (Gillies 1982, 68), which meant that such a number was to be defined only on the basis of arbitrarily adopted assumptions (axioms), which did not have to be grounded on any internal reason, neither did they have to refer to reality in particular. Their justification is only the proposed set of axioms, which needs to be internally consistent—it is basically the only absolute requirement. Mathematics based on such numbers was gradually becoming a formal system in the sense that it formed a system of interconnected and consistent formulas which did not have any interpretation other than them alone.

The manner of constructing mathematical beings as proposed by Peano certainly became the foundation of Hilbert's system and, in more general terms, a certain way of reasoning that made mathematics a science placed as if above and beyond the world. It also made it possible to use mathematics, especially numbers and arithmetic, to describe own mathematical beings. There was no contraindication against e.g. numbering mathematical theorems and perform calculations using the utilized numbers. The latter—higher—level was becoming a domain of mathematics. Peano's idea of number did not forbid it since it completely disregarded any obligatory, prior or objective interpretation of number. It did not point to any denotate of number, which meant it could be used in any way provided that this use did not violate certain pre-adopted formal assumptions. Of course, there could be many such sets of assumptions, with each of them becoming the foundation of separate mathematics—on the condition that these sets remained internally consistent.

The benefits of mathematics were made use of by said Gödel—he proved his revolutionary theorem to be true, which showed at the same time that the axiomatic system as proposed by Peano and Hlibert could not exist in the form envisaged by its creators. Gödel's concerns certain mathematical theories considered in very general terms and essentially proves that these theories cannot exist in an integral and complete form. Peano and Hilbert assumed that it was possible to construct such theories that would encompass the entirety of their own theorems, turning them into a complete, proven, and internally consistent system. But Gödel showed that it was impossible to arrive at such an entirety and that there would always be a sentence which would escape the procedures of inference. His pessimistic vision had quickly spread in a more or less correct form and affected the perception of mathematics itself, and even the nature of the world, as something falling fundamentally outside an exact and complete mathematical description (Krajewski 2003).

At the same time, Gödel's line of thought introduced a new technique of mathematical analysis, one that enabled the creation of a mathematical, higher-tier level of interpretation of mathematical statements, which became the starting point for e.g. its theorems (Nagel and Newman 1958). Speaking in common terms: Gödel numbered such theorems and then made calculations using the numbers assigned. Hilbert's program, mentioned earlier, was based on a similar idea involving the formalization of mathematical inference. The idea was simple: to substitute mathematical inference with pure mathematics—a bit more difficult and more abstract, but offering a certain exceptional possibility. This would make it possible to formally prove mathematical theorems. To verify mathematical theorems just like equations are solved: in a technically clear and consistent manner. And this was exactly what Hilbert's tenth problem from the list included in his grand mathematical program was about. The question underlying this problem was answered negatively by Turing in 1936 (the text was published in 1937, but written in 1936). Hilbert's idea collapsed, but what remained was a clever intellectual construct devised by Turing, which was applied later on in the process of proper inference. It was the already described idea which became the germ of the structure of a physical apparatus of a computer. An idea of formalization of mathematics, as well as a tremendous intellectual effort involved in the reconstruction of the manner of existence of number are what lies at the root of the theoretical concept that became the basis of the construction of a computer (Davis 2000; Krajewski 2003; Maciąg 2012).

The computer is a child of the problem of number, so to speak. What underlies the problem is the effort to understand the world in its most abstract domain. Ifrah, who approaches the history of the device very pragmatically, adding much more detail (Ifrah 2001), writes simply that the computer is a result of a very long historical process during which humans have been gradually perfecting their intellectual capabilities. But the story about the computer is also a story about the world and the conditions in which the computer knowledge has emerged and existed because it is based on a certain vision concerning all these elements. It is not a finite and uniform theory. Rather, we deal with an opening onto a vast and ever-infinite field of problems posed by mathematics.

The effort has not come to an end. It actually poses an increasingly greater challenge. At the end of the twentieth century, Alain Badiou made an attempt to ontologically interpret mathematics as a field of numbers. He starts his first meditation in his groundbreaking work entitled *L'Être et l'événement* (Being and Event) (Badiou 1988) from quoting the already mentioned idea of the One as proposed by Leibniz. The meditation concentrates on complicated dialectics combining the one and the multiple. Badiou claims that the world manifests itself as the multiple, which is contrary to the definition being offered by Leibniz. Badiou believes that what exists is only the multiple. The one is only a tool, an operation: "count-as-one". Every multiple is therefore always a multiple of multiples (Badiou 2006, 29). This way, the French philosopher constructs the main conditions of a general ontology. The theory of the multiple thus created—with associations with Cantor's work being highly relevant here—is, however, also the basis of mathematics, which reaches an unusual status this way; it finds itself at the highest tier of understanding of the world, determining the conditions of its own existence by itself. He claims that mathematicians, although generally unaware of this fact, have access to the level of reality, which has usually been the domain of philosophy only.

Badiou's metaphysical and hermetic reflection, while possibly highly inspiring—which will be useful when we analyze the multiple defined as a complexity, is challenging and difficult. The philosopher offers its more pragmatic face in *Le Nombre et les nombres* (Badiou 1990). He nominates a number, belonging absolutely to the field of the phenomena of multiples, to perform the realization of mathematics as an ontology: "Number is neither a trait of the concept, nor an operational fiction; neither an empirical given, nor a constitutive or transcendental category; neither a syntax, nor a language game, not even an abstraction from our idea of order. Number is *a form of Being*. More precisely, the numbers that we manipulate are only a tiny deduction from the infinite profusion of Being in Numbers" (Badiou 2008, 211).

The hermeticism of reference to being in this diagnosis is balanced by an extensive introduction where Badiou addresses the role of number in the world of today.

He lists the domains governed by numbers; these include: politics, so-called human sciences, cultural representations, economics. What is more, numbers speak to our souls as well. We live in a world where number is a despotic ruler. The paradox of the situation is, as Badiou claims, that we have no recent, active idea of what number is at our disposal (Badiou 2008, 1). The diagnosis leads to a dramatic conclusion: "we don't know what a number is, so we don't know what we are" (Badiou 2008, 3). The problem seems to be resolvable only at the level of the proposed ontological concept of mathematics and number.

The project created by Badiou, an atheist, a leftist, makes a critical return to Leibniz's deep intentions in the way that it retains the latter's metaphysical perspective. The notion of the metaphysical perspective is used here in its simplest sense, which elevates the reflection above the objects from the physical world and pursues a higher order assumed under this notion. Badiou abolishes the idea of the One, understood also as a substitute name of God. He substitutes it with the idea of the multiple instead. But the multiple is of a similar nature as the dethroned One: it is a being in itself, one that sets the conditions of functioning of the material world, leaving a felt mark in this world by means of its full potential. Material nature shall be understood here as, for instance, its historic nature, its entanglement in social matters, etc. Therefore, mathematics using e.g. a number, describing and understanding certain metaphysical conditions of the existence of the world, has insight into the ontic realizations of number in the reality that surrounds us. Badiou's work seems to offer the most advanced analysis of the context of existence of a computer—a mathematical machine—one can so far imagine, and highlights the potentially extremely deep sense and meaning of digital transformation.

Digitalization

So far, in the context of the considered notion of digital transformation, we have been trying, at least partially, to explain the significance of the problem of number, which exists as a symbolic construct—digit, but also as a sophisticated cognitive construct. In the latter case, it acts not only as a universal tool used to describe the world within such a field of knowledge as mathematics. Number can also be interpreted metaphysically, meaning that it can be considered as an autonomous, self-contained being, providing the foundation for other beings to exist. Such a presentiment, present already in the ancient times among e.g. Pythagoreans, appears as a general impression of experiencing the world which is naturally "mathematical".

This is how, for instance, Galileo describes the world in 1623 in his work entitled *Il Saggiatore* (*The Assayer*), using a metaphor comparing the universe to a book which "è scritto in lingua matematica, e i caratteri son triangoli, cerchi, ed altre figure geometriche"[3]—meaning that "is written in the language of mathematics, and its characters are triangles, circles, and other geometric figures" (Galilei et al. 1960, 184). The problem becomes a subject matter of intense deliberations leading to a range of sudden effects taking place at the turn of the 19th and 20th centuries; this is also when geometry appeared to be a significant source of inspiration. A highly advanced idea of ontology based on mathematical grounds was offered by Alain Badiou in the late twentieth century. This was covered in the previous chapter.

The described selected facts make a reality that escapes common knowledge and the much more disciplined scientific knowledge alike, especially if it deals with other domains of the world and simply has no time for an analysis of this kind. This does not mean, of course, that the underlying foundations of formation of the idea of numbers and the idea of mathematicity do not have an impact on the manifestations of reasoning, which affect them in some way. This impact, latent and indirect, is a separate subject, possible to be discovered by engaging in a reconstruction, archeology, as referred to by Foucault (1969, 1972), in attempts to recreate the course of discourses in history, creating states of knowledge locally in time and space. Therefore, according to the idea of historical placement, asking the question about the contemporary ways of making use of the notion of digitalization is necessary, which, in line with Foucault's thought, represent the evolving state of knowledge about digitalization. It needs to be stipulated, however, that the notion of "knowledge" here means only a certain state of conviction, not the state of truth, of reality as it is (Foucault 1971, 1981). Unless we consider truth in light of the pragmatist approach of William James, for whom it is only "the expedient in the way of our thinking" (James 1907, 150). According to this interpretation, truth is in no way grounded metaphysically, which makes it a universal foundation, but its nature is instrumental: it is a means, not an end. The process of the multiplication of truths, with each serving a different idea, is addressed by Bauman. Knowledge in the context of the lack of a single truth becomes a subject of social exchange and shapes the local image of the world for current needs (Bauman 1997).

The problem with the notion of 'digitality' is that it appears in two variants in literature: digitalization and digitization, which one has to come to terms with. The

[3] https://www.astrofilibresciani.it/Biblioteca_UAB/Biblioteca/Galileo-Saggiatore.pdf (accessed on: 10/22/2019).

said terms, like digital transformation, have been attracting a clearly growing interest since around 2014 according to Google Trends, which covers the period from 2004 onwards. The perspective of the recent fifteen years is not sufficient to perform a more extensive analysis of the trend but it does show that these terms are not overused at all. In an article featured in the encyclopedia on the theory of communication and philosophy, Brennen and Kreis mention these notions being used interchangeably in practice and suggest a clear distinction between their semantic areas, governed by *The Oxford English Dictionary*. They describe digitization "as the material process of converting analog streams of information into digital bits. In contrast, we refer to digitalization as the way many domains of social life are restructured around digital communication and media infrastructures" (Brennen and Kreiss 2016, 556).

In both cases we are dealing with certain processes that cause change. However, in the case of digitization, what changes (transforms) directly is an object, and in the case of digitalization, what also changes is the environment surrounding a given object that has changed. Digitized objects cause a further digitalization of the world. A technical process makes the contents of the world change and objects become accompanied by their digital copies, which also changes the world itself.

Brennen and Kreiss underline two aspects of this technical process, which appears to be the source and the reason of further changes: the symbolic aspect and the material aspect. The former means replacing "analog signals" with bits, expressed as zeros and ones. This kind of symbolic standardization and the simplicity thereof make anything that is digitized expressible using very different—and real—means, provided that they maintain the fundamental dichotomous form. It does not matter what material this form is expressed in, although in the case of computers, this material is only alternate electric current. But one can have certain reservations regarding the interpretation of this notion, apparently somewhat simplified. It is possible to digitize much more than just signals. There is an actually ongoing race for an extension of the domain of digitization, and subjecting more and more images, artifacts of various kinds, and much grander things—like buildings or celestial bodies—to this process is nothing new, in fact.

One of the domains where this race takes a particularly intense form is the domain of language. It is also probably the most advanced such domain in terms of the object of digitization—and this object is the semantic content of text. It is an ambitious task, aimed at overcoming the fundamental limitations of the computer, which—in its standard form—'sees' text only as a set of certain graphic representations. These representations may reveal some similarities and patterns at the syntactic level, but not much more. Many attempts to overcome this limitation, meaning to equip computers with some very basic procedures imitating human cognitive abilities, have been carried out in research on the so-called artificial intelligence, conducted since the 1950s. One of the tasks leading to the establishment of such procedures is the problem of digital representation of knowledge in a form enabling automatic reasoning. The first formal models of knowledge appeared in the late 1960s and early 1970s, and the later course of development of this field is quite turbulent (Flasiński 2016; Müller 2018; Russell and Norvig 2010; Warwick 2011).

A relatively new invention in the area of digitization of semantics (Mikolov et al. 2013), or more precisely: in the area of artificial intelligence involving natural language processing (NLP), which has found a growing range of very promising application scenarios, is about assigning series of numerical values to words, representing a given word's position in the semantic space with respect to other words. This way, we arrive at specific semantic "maps" (or, more precisely: spaces), which can be compared and analyzed.

Semantic representations of words assume a mathematical form of vectors, meaning ordered lists of numbers based on which one can perform mathematical operations such as addition or subtraction. The basis for the creation of such representations is the assumption that words close in meaning tend to 'cluster' near each other, and the larger the body of texts based on which the analysis of the position of words with respect to each other, the greater the proximity of their semantic content. Operations performed on vectors representing the semantic content of words convey their semantic value. For instance, "queen" minus "crown" equals "woman", which may not be the greatest achievement in the field of hermeneutics but it does surpass some level unattainable thus far.

For Brennen and Kreiss, digitalization is a process much broader and more multifaceted compared to digitization. They remind that the term was first used by Robert Wachal in his essay entitled *Humanities and Computers: A Personal View*, published in the "North American Review" in 1971 (Wachal 1971). The text initiated, as they believe, an extensive discussion, which they prove by quoting names like Castells, van Dijk, Machlup, Bell, and others. They are a complete reflection of the social phenomena taking place for a long period of time. The former involves rapid cultural and economic changes of the late 1960s and early 1970s, when publications by Peter Drucker (1961), Fritz Machlup (1962) or Daniel Bell (1973) described the sudden social and economic paradigmatic transformations. A certain breakthrough arrives with the idea of the information society emerging in the early 1980s (Masuda 1981). In his book, Masuda presents the world as dominated by computers connected with each other by means of a network, and calls this creation a *computopia*. Manuel Castells and Jan van Dijk published their most important books in the late 1990s, when the Internet was at a stage of much welcome and celebrated development and at the same time brought to light the phenomenon of a network, which soon became the most general image of how the world was organized (Castells 1996; van Dijk 1999).

Digitalization as a social and economic process, where the computer becomes a tool utilized in a growing number of spheres of life and economy, can be seen well in the 1980s. "Electronics" magazine was launched in 1973. Its catchphrase was "The Great Takeover", and the content was about covering the increasing number of digital solutions replacing their analog predecessors in a range of devices. But it was only ten years later, in 1983, when "Time" magazine announced the computer as the "Machine of the Year", with a headline reading *The Computer Moves In* (Ceruzzi 2003, 346). Masuda published his visionary book some time earlier, including his observations of the IT industry in the early 1980s in Japan, which far outstripped other countries in this area. In the 1990s, the general fascination shifts more and more onto the increasingly common and potent Internet as described by Castells and van Dijk. Computers connected into a network appear to be a source of new social forms of human presence in the world, redefining its nature (Papacharissi 2010; Turkle 1995). The world of network relationships and dependencies emerges, changing, basically, all aspects of the social life and culture as known up to then. The initial idea of digitalization, understood as the development of computer technology, thus grows to become a separate and very event-rich story.

Schalmo and Williams ground their discussion on the idea of Brennen and Kreiss to interpret the existing alternation of digitization versus digitalization, but they clearly change the perspective at the same time (Schallmo and Williams 2018). Their book, published in 2018, focuses on digital transformation described from the point of view explained well by the subtitle: *Guiding the Successful Digitalization of Your Business Model*. This is a characteristic type of approach to this issue, which has been taking shape in recent years: digital transformation emerges as a problem affecting mainly economic matters, especially in the area of management. Brennen and Kreiss defined digitalization in a completely different context: the context of the theory of communication and philosophy, so they were not concerned with the pragmatics related to rational economic activity and its essential requirements, mainly concerning its clear purpose: profit, effectiveness, etc.

Schallmo and Williams draw examples and relate them to sources almost entirely focused on business. They are mainly interested in one aspect of the alternation in question: digitization. This stems from the answer they give to a question they pose themselves: "What is digital?". The answer is a description of the reality of corporate management in the context of digital transformation, which is simply a story of "how companies run their business" (Schallmo and Williams 2018, 3). They cite several definitions of digitization, including e.g. a conversion of the analog into the digital, a mediation between the material (e.g. signal-generating sensors) and the immaterial (e.g. business processes), or a unification of different entities such as people, data, processes, and things within a standardized business process. They themselves define digitization as a digital way of incorporating analog or physical artifacts into a business process, which means approaching the process instrumentally, theologically, and purely practically, with the nature of the process alone being quite banal: technical and functional.

Schallmo and Williams discuss the process of digitalization across only a dozen or so lines of their text. The context remains business, for which digitalization is a description of changes in its operations and models, with the changes resulting from knowledge acquired through digital initiatives, creating added value in an enterprise. They arrive this way at a bit higher level of abstract, involving shifting from the level of objects to the level of knowledge. The domain of reality stays very narrow and purely pragmatic, though, and its limits are marked by the boundaries of an enterprise.

Schreckling and Steiger explore the same business reality and understand the 'digitize/digitalize' alternation in a similar fashion (Schreckling and Steiger 2017). Their book contains an in-depth analysis of both processes taking place in the world of corporate management. The pragmatics of their discussion is therefore clear and completely subordinate to the narrow purpose of business. They repeat their definition of digitization after Gartner: it is simply a process of changing the analog into the digital. They see the effects of this process in much broader terms. They argue that it significantly affects the economy and society, both as certain parts and as a range of dimensions, which come from the sphere of the economy anyway. These dimensions include business models, consumption, employment, investments, etc. Like Gartner, they view digitalization in a narrow and specific way, as the use of digital technologies in order to modify business models. Schreckling and Steiger analyze and systematize both processes in a disciplined manner, without moving outside the corporate reality.

Legner et al. (2017) see the 'digitization/digitalization' alternation as a mere introduction to further issues, which simply need to be brought to light due to the number of misunderstandings and misconceptions. Like their predecessors, they define the first term only as a technical process of converting analog signals into digital signals. Its 'dematerializational' nature exposes it greatly to further manipulation, which does not have to mean—let us note—only purely technical consequences, as authors argue. They place an emphasis on the second notion, digitalization, which they define very broadly and which means: "the manifold sociotechnical phenomena and processes of adopting and using these technologies in broader individual, organizational, and societal contexts" (Legner et al. 2017, 301).

The latter of the discussed texts are clearly governed by an imperative of a focus on detail and precision since the process in which digitality appears in the world seems to be increasingly complex and common, so it needs better, more precise notions to be described. And this is probably how the alternation in question is born. Earlier texts, like the book by Vogelsang, did not take this alternation into account at all even though they offered a similar economy-oriented approach (Vogelsang 2010). Vogelsang does not use the term 'digitization'. And he is not the only one. Apart from the publications that make attempts to deal with it and match it with the older notion of digitalization, like the already mentioned texts, there are also works exploring only one of the said notions. Such an approach may lead to both a focus on technology as the main factor of change in digital transformation and an almost complete abandonment thereof, with the context of business and enterprise still remaining present.

Olleros and Zhegu choose the notion of digitization, which they see as "the full range of software-driven processes – all the way from datafication and computation to prediction, display, communication and action – that allow increasingly smart machines to intervene in the world" (Olleros and Zhegu 2016, 2). Their understanding of the process is purely functional and does not refer to existential criteria, or even to ontological criteria resulting in such an alternation as 'digitalization' *versus* 'digitization'. They are not interested in the essence of the process but in its actual materialization. The process is materialized in a universal tool—a computer, which diversifies its operations and makes them more specialized, meaning it becomes function-oriented and need-oriented thanks to its software. This is enough to create a definition of and diagnose the functioning of the process in the world, characterized by an unprecedented level of ubiquity and impact. These qualities make it possible to consider the process of one of the so-called GPT—General Purpose Technologies such as the steam engine or electricity (Bresnahan and Trajtenberg 1995).

According to Olleros and Zhegu, many qualities of digitization make it reasonable to consider this technology a force even more influential and powerful in its ability to reshape the world than, for instance, the said electricity. These qualities include continued development and changeability, the ability to transform existing possibilities and revealing new ones, enhancing the synergy in business processes, the introduction of new variants of man-machine cooperation, reorganization of fundamental economic mechanisms—involved e.g. in making profit, etc. When listing these qualities, Ollerus and Zhegu focus mostly on the areas of economics and economic management, thus leaning towards a previously obvious and dominant monotony of reflection on digitalization, which appears in their text in the form of the newer notion of digitization. The field in which it is used does not leave much choice, in fact. This monotony, this one-sidedness of the said reflection is surely a trait of dominance, or even the hegemony of their economic discourse.

But the question which has already been asked here and seems to be simple and straight-to-the-point, "What is digital?", does not have to entail an answer that makes use of any of the said notions: digitization or digitalization. Interestingly enough, such a reflection tears the narrow walls of the field of management virtually instantly. Such a question is asked by Francesco Varanini (2018) in a text included in a collection on management, and, speaking more precisely, on: acting as a manager in an enterprise intent on taking advantage of the digital transformation (Bongiorno et al. 2018), which still fits in the dominant trend. But his answer is completely different. It starts from recalling the once very influential and famous book by Nicholas Negroponte, *Being Digital* (Negroponte 1995). The book was published and translated into dozens of languages at the early stage of development of the Internet, right before its dynamic evolution and the equally rapid, enthusiastic increase in the expectations and hopes invested in it.

Negroponte based his text on the observations of Marshall McLuhan, a legendary researcher of the media and the civilization subject to their influence. This kind of field of observation is slightly different from the one referred to in previous texts, corresponding to the article by Brennen and Kriess quoted in this book. However, he does not get tangled in notional innovations that try to overcome the challenge, introducing simple classifications and alternations such as 'digitization' versus 'digitalization'. Varanini takes the problem into the area of the humanities understood in broad terms, where digitality is seen as a new reality which is born and matures accompanied by digital technology, engaging in numerous complicated relationships with it. Its followers are no experts in technology, but they notice its presence, and the most important sphere where it becomes a dominant axis of the world is human, of course.

One can hazard a hypothesis that the notion of digitalization, also in the newer version of the notional alternation of digitization/digitalization, takes the thought necessarily through the territory of technology, where the traveler faces various species of fierce monsters breathing the deadly fire of engineering, IT, and even mathematical knowledge. This is, of course, a result of the researcher's temper, recalling the discussion that took place among three great philosophers in the late nineteenth century: William Dilthey, Wilhelm Windelband, and Heinrich Rickert. It concerned the explanation and legitimization of the coexistence of two completely different manners of describing the world in scientific terms. The discussion dates back to much earlier times, though. To the age of the disputes from the second half of the eighteenth century, which sought to question the achievements of Newton and Leibniz. The best-known party thereto was Immanuel Kant, who inspired the later three by providing them with essential notions and topics for discussion (Kant 2004 [1786]).

Dilthey is the author of the contemporary interpretation of the area of *Geisteswissenschaften*, encompassing what we presently refer to as "the humanities and social sciences", meaning disciplines such as psychology, anthropology, political economics, law, history, language and literature studies, aesthetics, and philosophy (Makkreel 1973, 37). It is reasonable to stress here, as recalled by the quoted Makkreel, that the translation of this German word into English is highly problematic. This is rather governed by a certain interpretation and circumlocution, which are there because of the absence of the right equivalent in the context of the German notion of *Geist* (spirit), co-shaped to a great extent by Hegel. In the English language, the word *spirit* has a much narrower and more specific meaning. In the Polish language, the problem is similar; although the notion of spirit is still present and well-rooted, it has lost much of its significance and common nature today.

Dilthey's main objective was to prove the methodological difference between *Geisteswissenschaften* and *Naturwisenschaften*—the latter encompassing strictly empirical sciences, such as physics or biology. Secondly, thanks to this difference and autonomy, Dilthey tried to prove the separate and equivalent nature of the former: "the justification for this methodological claim central to Dilthe's delimitation of the Gesisteswissenschaften is to be found in epistemological terms. It is philosophy's task to provide an epistemology that can show that the *Geisteswissenschaften*, although not as clearly definable in their first principles as the *Naturwisenschaften*, are not less fundamental, comprehensive, and objective in their results" (Makkreel 1973, 38).

In the 1950s, Dilthey's opposition was restored by Charles P. Snow in a lecture on the dramatic divide of the world of science into two cultures: humanists and scientists. The lecture was later published in an extended version, continuing to persistently exist in the discourse on science (Snow 1962).

Despite the passage of time and the tempestuous transformations affecting the domain of science, meaning the ways of understanding, ordering, and justifying the descriptions and propositions it has produced, Dilthey's old opposition, existing in a slightly trivialized version offered by Snow, continues to fare well in the context of the problem of digitalization. It is still possible to name fundamental and opposing approaches that seem to resemble it. One is a pragmatic, solution-oriented approach, one well-grounded empirically—referred to as an engineering or managerial approach. The other makes good use of a humanist, contemplative nature, full of careful presumptions expressed by means of a complex notional apparatus which is able to handle such challenging categories as human or society. But it does not mean that there is no deeper reflection which transcends this historic, obsolete division. Its continued existence may, however, bear testimony to the primal, simple, and introductory nature of the dominant reflection on digitalization. This means that it poses an unsolved mystery to the reflection in question, and that the first steps into the understanding of the complexity of the phenomena underlying it must be taken elsewhere.

What may seem a paradox, traces of such a mature approach appear in relatively early times, meaning during an increase in the interest therein in the second decade of the twenty-first century. After all, 2010 was the year in which Peter Vogelsang's book on digitalization, 'inaugurating' the period in question, was published (Vogelsang 2010). Vogelsang takes his argumentation through the plane of economics, which means he follows the dominant trend of reflection on digital transformation. Yet, at the beginning, he gives some space to the understanding of a phenomenon whose history is described by starting out from the invention of Gottfried Wilhelm Leibniz, depicted as the creator of the binary system. To him, the history of digitalization as a technology does not start with the emergence of the computer, but with the arrival of the telegraph, which came into being in the early nineteenth century and connected all continents in the second half of that century. IT, which takes over and pushes the telegraph's potential further, is considered by Vogelsang as well as Olleros and Zhegu alike a GPT (General Purpose Technology), the existence of which is based on general social issues. In this context, Vogelsang refers to the idea of an information society.

Vogelsang represents an advanced manner of describing digitalization, placing it in various interpretation contexts—including those philosophical, careful and sparse, but essential to the further understanding of the idea. This approach is not developed much in the further reflection on digitalization. It is replaced by a simplified classification, based on trivial, practical observations, which can be surely useful in operating a business. This kind of pragmatics can be clearly seen dominant in the area of reflection on digital transformation. In the light of this pragmatics, digitalization seems to need no detailed description. An example of such an extreme approach is an argument by Scheer (2017). One of the few examples of another, more insightful understanding of the notion of digitalization, but still remaining within the domain of economy-oriented reflection, is a book by Jan Christoph Bartodziej (2017). The topic it addresses, however, is an economy-related one, i.e. the concept of Industry 4.0, meaning: "a strategic initiative", launched in 2011 by German government, combining technical, economic, and socio-political factors to improve Germany's global competitiveness. The general and basic nature of the concept of Industry 4.0, its political and social essence, make Bartodziej depict digitalization, one of its fundamental processes, equally seriously and from a long-term perspective. He views it as a step towards a digital revolution which will cause a "social change of an industrialized society to a post-industrialized, knowledge-based, service-oriented, information-based society" (Bartodziej 2017, 33).

A similar intention can be found in a book edited by Ulrich Sendler (2018), discussing the phenomenon of the Internet of Things—a technological solution on the one hand and a promising business project on the other. In its case, like in the case of many other technologies, this alternation is virtually inseparable in everyday practice. Like Bartodziej, Sendler sees the Internet of Things as a part of the German economic project dubbed Industry 4.0. He understands digitalization as a "megatrend"—a top-priority process, one of fundamental significance to the future of societies and humans, and one which also involves some serious threats. It is an interesting vision, portraying digitalization almost like a natural element. Its autonomy and dynamism, implying its uncontrollability, and the promises it makes at the same time make it quite special. Sandler's approach reveals a romantic overtone of railway or bridge builders, heroic explorers of the world, making it civilized through their efforts. Today, such a description, in the face of the dreadful effects of human activity in the period of the Anthropocene, cleared forests, contaminated rivers, and changing climate, must sound ironic to say the least.

The collection edited by Sendler also features a text by Martin Eigner, introducing one more—different—type of use of the term 'digitalization', referring to its purely technical meaning (Eigner 2018). The background is the said Internet of Things, important for a very trivial reason: it is an area of economy of great potential. Eigner quotes some numbers. By 2020, the number of devices interconnected within the Internet of Things will reach 37 billion, and the value of the assets invested in the process will amount to around USD 5 billion. The Statista website offers more accurate data. According to the company's estimates, the IoT market will reach a value of about USD 25 billion in 2020 and will continue to grow at an almost exponential rate to be valued at an astounding sum of USD 1.5 trillion in 2025. The number of IoT connected devices in 2025 will exceed 75 billion, after a period of rapid growth in preceding years.[4] This amazing growth and the huge numbers can all be achieved as there are sufficiently efficient ways to design them and make them a reality, based on digital prototypes. Digitalization—thanks to the transfer of the biggest part of the process of creation of the said prototypes to the digital domain—will simplify creation, facilitate construction, and enable collaboration. Dear reader, welcome to the world of Industry 4.0, a world of manufacturing processes gone digital, thus becoming more flexible, more efficient, and less expensive. The great level of expectations accompanying this vision is hardly surprising.

The enthusiastic and affirmative attitudes, seen in the writing of both Bartodziej and Sendler, are, in fact, a standard in literature addressing the issue of digitalization, being an important part of digital transformation understood as both a business and an economic process. Even when it mounts great challenges to businesses, which is raised frequently in the said literature, it is still considered a great promise of improved performance and growing profits. Digital transformation and digitalization tend to be eagerly placed in the pragmatics of management on the side of advantages that appear immediately after certain essential organizational measures are taken.

[4] https://www.statista.com/topics/2637/internet-of-things (accessed on: 10/24/2019).

This optimism is moderated and balanced by critical literature, which goes beyond the trivial business pragmatism and tries to analyze the process of digitalization in a complete social context, with particular consideration of human good and social justice. Horkheimer—considered the main lawmaker of the Frankfurt School's program—has described the approach as follows: "Critical thinking (…) is motivated today by the effort really to transcend the tension and to abolish the opposition between the individual's purposefulness, spontaneity, and rationality, and those work-process relationships on which society is built. Critical thought has a concept of man as in conflict with himself until this opposition is removed. If activity governed by reason is proper to man, then existent social practice, which forms the individual's life down to its least details, is inhuman, and this inhumanity affects everything that goes on in the society." (Horkheimer 2002, 210 [1937]).

In the quoted passage, Horkheimer refers to the fundamental opposition of the human world, to the tension emerging between a human, a being protected in its uniqueness and freedom, fragile and suffering, and the entirety of other similar beings, in which they lose their individualism in favor of a pragmatic, organized, purpose-oriented community referred to as a society. He warns, in particular, against excessive pragmatism, extreme organization, or exorbitant purposefulness, which go hand in hand with overregulation of the principles of existence of the said community. It then becomes inhuman because it leads to depreciation of the said uniqueness and freedom. It is a thinking deeply rooted in the Western model of humanity, with its final form designed by Kant (2004 [1786]). The model in question was an extraordinary, unique attempt to discuss, justify, and protect the human persona, shaped by the European history and culture, but—hence the past tense—it faced a fierce defiance in the twentieth century, also as a result of a new civilization being formed under the influence of the Internet and IT (Maciąg 2016).

Horkheimer does not stop at demands for and wishful thinking of a better world, built on a harmony connecting single individuals with their social environment. On the contrary. He also defines the hard duty of the critical theory, being the immersion in the reality of the current relationships determining the everyday life. Among these relationships, he focuses, in particular, on those affected by all the factors that disturb the said harmony, exposing humans in their entire weakness and defenselessness to harm. It's a simple outcome of a consistent further step in reasoning: a certain responsibility and duty, a kind of moral obligation. It implies that writing about humans and their situation may not stop before the *concretum* of their everyday, ordinary existence in which all forces forming the pyramid of their more or less theoretical social relationships are materialized.

The materiality of social relationships, the entanglement in the current historical time frame, the local dependencies on others suggest Marxist tools of description, mainly the dependency on the system of circulation and distribution of goods, determining the social position of each individual with absolute accuracy. This way, we arrive at one of the most important issues, which is power, and the reflection becomes inadvertently political to some extent. At the same time, the said tools channel our thought towards a historical synthesis, and historicity understood as a single fate also takes shape as a description of the process of social events arranged in a historical panorama. This sort of synthesis leads the observer of the modern world, Horkheimer, to highly pessimistic and prophetic conclusions: "(…) the critical theory of society is, in its totality, the unfolding of a single existential judgment. To put it in broad terms, the theory says that the basic form of the historically given commodity economy on which modern history rests contains in itself the internal and external tensions of the modern era; it generates these tensions over and over again in an increasingly heightened form; and after a period of progress, development of human powers, and emancipation for the individual, after an enormous extension of human control over nature, it finally hinders further development and drives humanity into a new barbarism." (Horkheimer 2002, 227).

Concern, pessimism, and—eventually disapproval are all unquestionable elements of critical thinking, which is clearly becoming a kind of political activity as well. But it is also necessarily present in the previously discussed affirmative approach, taking the more or less hidden enthusiasm around digitalization, which is the main subject here. Such type of approach becomes perfectly visible when digital transformation is discussed.

Franklin follows Horkheimer's assumptions in an almost textbook form, making the notion of control the axis of his description of digitalization (Franklin 2015). The basis of a very general understanding of the notion in question is offered by the idea of Frederick Jameson, covered in a well-known book entitled *Postmodernism, or, The Cultural Logic of Late Capitalism* (Jameson 1992). Cultural logic is understood there as a hegemonic, systematic cultural norm which lies hidden on the bottom of cultural production, thus shaping and molding it in a highly diversified manner. The work is not banal. Quite the contrary, it is deep, reaching for the ways in which humans understand the world, tell about it to each other, and pass judgments. The three seemingly simple acts open, in fact, the door to great and interconnected theoretical domains such as epistemology, culture, or ethics. Jameson seems to believe that the logic of late capitalism—this political and economic context is, as we know, natural to critical reflection—is subject to the rules of postmodernism, which also calls for an appropriate interpretation of the latter notion.

Franklin substitutes the notion of postmodernism with the notion of control. He understands it most of all as a construct addressing the way of human understanding of presence in the world. This presence is undergoing a significant transformation, and its sources are digital technologies, information processing systems, which "reconceptualize" the social reality. The control referred to by Franklin "describes a set of technical principles having to do with self-regulation, distribution, and statistical forecasting that is extended to the conceptualization of sociality through a series of subtle historical transformations" (Franklin 2015, xv). He underlines that it manifests itself through various practices and objects, "from cybernetics to economic theories and management styles, to concepts of language and subjectivity, to literary texts, films, and video games" because it fulfills the role of a basic epistemological assumption. It is a universal rule of existence, which functions among other characteristics, typical of every fact of worship, bringing each time an own specificity to the table and making such a fact unique and common at the same time.

The key to the understanding of control as viewed by Franklin is the notion of digitality—Franklin understands it in an advanced and interesting way. First, he transcends its material denotation in the form of technologies of various kind. It seems that although he does not put it in a straightforward manner, he sees digitality not as a presence in the world of technological inventions, which affect everything they are applied with because of their unique qualities, but as a feature of things the world is made up of and thus impart a given quality on it. This considerably increases the set of objects, not necessarily related directly to digital technology. A quality common to all of them is the state of being, which is not continuous, but quantified, discrete: "Digitality, as it functions under the sign of control, comes to describe not only a set of technologies or logical operations but also a fundamental condition. To exist, from the point of view of control, is to be digital (…). This ontological digitality, separated from the machines and interfaces with which it has become synonymous, entails a fundamental process of discretization that can be purely conceptual as much as it can enable particular technological processes. In the same way, digitality moves from a specific system of representation and technical processing to a set of generalized metaphors" (Franklin 2015, xix).

Franklin's reflection, most interesting from the point of view of this text, is connected with the notion of metaphor he has introduced. The notion is based on an epistemological distance to reality. As claimed by Lakoff and Johnson, metaphors are the foundation of the conceptual system, meaning a certain construct, specific to each individual, used to understand the world and serving as the basis to make decisions and act (Lakoff and Johnson 1980). By virtue of its nature, a metaphor is local, particular, and exceptional. It does not standardize nor generalize. The implicit assumption underlying the idea of a metaphor is the existence of many very different views of the world, some of which may become dominant. The epistemological perspective in the case of metaphor becomes dispersed, blurred, lacking a solid ground. Such a vision of the world loses its sharpness and basis. It may also become manipulated, or even falsified: to influence people and the world. And according to Franklin, the metaphor of digitality does have such power.

Digitality as a source of impact on people, things, and societies is the subject of a more extensive collection of essays by many authors, edited by David Chandler and Christian Fuchs—*Digital Objects, Digital Subjects: Interdisciplinary Perspectives on Capitalism, Labour and Politics in the Age of Big Data* (Chandler and Fuchs 2019). The book, like Franklin's work from some years before, is also consistent with the broad domain of critical reflection. It differs, though, in the way the reflection is understood and used. Christian Fuchs is an author of many works where critical reflection is employed as an interpretation basis (Fuchs 2008, 2012, 2015, 2016). His approach in particular emphasizes inspirations drawn from Marx, which is also clearly visible in the field outlined for the book in question. A few years ago, Fuchs, with his customary orderliness and scrupulosity, proposed the "dimensions" determining the space of subjects of the critical theory of society, which included: critical ethics, critique of domination, exploitation, and alienation, dialectical reasoning, ideology critique, and political struggles being a materialization of political practice (Fuchs 2016, 6). This list, obviously based on Marx's thought, is simultaneously a program of a matter-of-fact analysis in the area of political economics.

The three problem areas of the book in question are the classical subjects of critical social analysis, conducted in a spirit of the said economics, supplemented with the adjective 'digital', resulting in, accordingly: digital capitalism, digital labor, and digital politics. Fuchs and Chandler's historical materialism forces a vision of the world upon them where the computer, having seized all spheres of life, creates a new political system in a deterministic and real fashion (Chandler and Fuchs 2019). Moreover, Fuchs and Chandler point to a particular technology that comes to hold a special and dominant status among others, to eventually become equated with digitality. This is to mean *Big Data*—a technology, an organizational or business strategy, or, finally, a certain way of understanding and viewing the world, reaching a theoretical epistemological level. The term "digital capitalism" is used interchangeably with "Big Data capitalism". The authors spare no effort to explain why the meaning of this notion—appearing in various contexts, starting from philosophy or theory of knowledge ending with social or organizational *praxis* in many variants. They also recall Leibniz and his binary system, which proves the relatively distant beginnings of "digital logic", whose shadow has lingered over social life since a relatively long time.

This level of generality, typical of Fuchs' writing, the problematization of the observed phenomena, the thematization of the noticed trends, processes, and facts are all in opposition to the material and historical reality: "[Computer] reduces life to calculability: zeros, ones and quantities. Computer algorithms cannot understand love, ethics, morals, solidarity, care, effects, and other distinct human qualities. The reduction of the human to computing and instrumental reason therefore also brings with it certain dangers, such as the application of digital machines for exploitation and domination. At the same time, the human use of computing also advances new potentials for cooperation, community, solidarity, resistance and sociality that, even from within capitalism, point beyond it. The computer within capitalism is therefore the antagonistic digital machine. Digital and *Big Data* capitalism is an antagonistic societal formation that deepens alienation and exploitation while at the same time advancing potentials for liberation." (Chandler and Fuchs 2019, 15).

This critical and Marxist approach results in an interesting description, equally practical and concrete as its inspiration, but still inconclusive. What is more, its materialism eventually trivializes the problem. The computer as the central figure of social and political reality becomes disappointing when it loses its metaphorical value and turns into a box full of wires. Especially the quoted claim of its inability to experience emotions or values is too much of a banal simplification to explain the complex nature of its presence in the world. The proof of this complexity is the said inconclusivity, which sees the considered device as a tool of both enslavement and liberation. To understand this paradox, we need a slightly different, more general, higher-tier description, where the ambiguity disappears, but not because of the applied dialectics, but thanks to a new understanding. And this tier of description is provided by Luciano Floridi's idea of re-ontologization.

Luciano Floridi is a prominent researcher studying the issue of information belonging to the world of IT—Information Technologies. He has been conducting his studies for over twenty years, constructing theoretical grounds of description and analysis based on two fundamental research perspectives: the philosophical perspective and the sociological perspective. He has published many books on the matter, with three monographs considered the absolute lead: *The Ethics of Information*, *The Philosophy of Information* and *The Logic of Information* (Floridi 2010, 2013, 2019). These works make up a perfect whole with a book portraying the tale of information technology in a metaphor of revolution the human world is going through as a result of its impact: *The Fourth Revolution: How the Infosphere is Reshaping Human Reality* (Floridi 2014). This fourth revolution takes place under the auspices of Alan Mathison Turing, who created the conceptual foundations of the computer. The father of the three previous revolutions were—accordingly: Copernicus, Darwin, and Freud. The latter invented the idea of these revolutions to stress the significance of his own studies. Freud presented this construct in his eighteenth lecture from a series delivered in 1915–1917 and published later. This is described by, among others, Bruce Mazlich, the author of a book with a similar concept to that adopted by Floridi based on Freud's output. The computer is portrayed as an invention causing rapid transformations in the history of mankind (Mazlish 1995).

It is easy to get confused when faced with all these metaphors. Floridi's book and concept should not be mistaken for e.g. the idea of the fourth revolution as proposed by Klaus Schwab, covered in his books from 2016 and 2018 (Schwab 2016; Schwab and Davies 2018). The latter revolution is coincident with Floridi's ideas only in the number, and concerns many rapid changes known under the name of industrial revolutions. Schwab, also the founder and president of a very significant conference of the World Economic Forum—taking place each year in Davos—represents a pragmatic and business approach, which we have come across in the context of digital technology and will return to. He stands in opposition to the reflection of Floridi, who is very contemplative, detached, viewing the world as a highly complex work of human creation, not just a market.

According to Floridi, the fourth revolution involves depriving humans of another, perhaps the last implication of their special position in the world, or even in the universe. Earlier revolutions have stripped humans from the sense of cosmological specialty and central position (Copernicus), from the sense of the uniqueness of being, aligning them with other animals (Darwin), and deprived them of the authority and power over themselves, which they apparently enjoyed thanks to free will, subjected to the dark powers of unconsciousness (Freud). The fourth act in the process of reconstruction (or: deconstruction) of identification of the role and position of human took place in the context and because of Turing's invention—a machine able to take over highly sensitive competencies of humans. Floridi defines them as logical reasoning, information processing, and intelligent acting. He moves on to describe human as the only current master of the world of information—the infosphere, where unique human abilities have been employed to manage it and multiply its resources. These resources include the entire human legacy: civilizations, culture, science, social norms, languages, stories, arts, music, poetry, or philosophy, which form a soil accumulating for years thanks to painstaking efforts of whole generations, upon which new creations tend to bloom. This exceptional position becomes discarded and the said competencies handed over to machines, which Floridi refers to in short, collectively as ICT (Information and Computer Technology).

Floridi's vision is founded on many notions which he uses to describe the reality as it becomes under the influence of the advancement in digital technologies. The first of these notions, mentioned earlier, is infosphere. It was coined in the 1970s, in imitation of the term "biosphere". According to Floridi, it has two meanings: a narrow one, describing a certain limited environment of all information beings and their relationships, and a broad one, according to which an infosphere is a synonym of reality per se, viewed from an information perspective. In the former case, an infosphere is a kind of an experimental fragment of the world. In the latter, it is an entire world that also exists– and, thanks to the development of ICT, most of all—as a being subject to the laws of information, governed and molded by information, and offering no way to escape it. An infosphere in its second, more general meaning is inhabited by inforgs. The name is short for *informational organisms*, beings whose existence happens in an environment of information, feeding on information, producing, and utilizing information.

Infosphere can be imagined. Actually, strictly speaking, it exists in a highly advanced form. The notion does not name an abstract vision, but a very real thing. It exists under the commercial and technical name of *Big Data*. Its major tool is the so-called Internet of Things. The term was coined in 1999 by Kevin Ashton as a reference to an electronic network of interconnected real and virtual objects. Each of these objects need to feature an own system of control and data collection. This way, it can become either a transmitter or a receiver of information. In the 2020s, the number of these objects, as already mentioned, will be astounding, reaching dozens of billions. The name 'object' is highly general and does not impose any limitations. The data these objects will be able to exchange can also be of any kind, provided that it is possible to express them in some numerical form. This concerns the more obvious information, like the information about the condition of a car engine during driving, and the more exotic information, like the chemical composition of bananas on a farm managed under a specific political system. The imagination is unlimited. Moreover, we are witnessing a certain race: various entities competing in inventing new kinds of data which—when interpreted correctly in cross comparisons with other data—will produce such data that will provide answers to the most complicated or difficult questions. Changing the location and information provided by a car engine may be used by an insurance agent to assess the risk related to the driver's habits. The chemical composition of the banana pulp in South America combined with our weather data and voting results can help producers of fruit preserves make the right decisions, and so on. Let us imagine a world consisting only of information transmitters and receivers, with data streams and rivers flowing and crossing between them all the time. Since this information is very monotonous in its shape—it is only zeros and ones, such a world is, basically, monochromatic. And this is exactly the essence of infosphere.

Its natural inhabitant is an *inforgs*—one of the objects producing and consuming information. Floridi uses this notion mainly to define the new role of the human, who—dethroned and stripped of their special nature through the revolution described by the philosopher—becomes equal to things. It is a dramatic act following a reversed progress, which now—unlike in the case of the European philosophy of the ancient age—involves the depreciation of the human persona. The great scheme of the apologetic and affirmative idea of the human persona, developed consciously in Europe since the Renaissance, expressed clearly by Descartes and eventually summarized by Kant, experiences a heavy blow. And this is included among the other impacts brought by the twentieth century, whose origins date back even to the nineteenth century. The latter belong to a phenomenon that leads to a negative attitude called antihumanism, and then—once the abolished scheme is decomposed further—to posthumanism. Both are based on a defiance against the source, historical design of humanity, which—like other designs—is only a grand illusion, calling for improvements as the world changes. Without them, it may disappear just like it appeared since, as concluded by Focault in his *The Order of Things*, its nature is like that of a trace on a sand—fragile, sensitive, vulnerable. The described evolution of perception of mankind has far-reaching consequences, which I attempted to describe in my book on the civilization of the Internet (Maciąg 2016). Technology is consistent with the above process, weakening the human's position, lowering their rank, depriving them of their privileged and exceptional position, eventually supporting the pulverization of the old European humanistic scheme.

Another term from Floridi's vocabulary, which refers strictly to the human on their dramatic path to the role of an inforg, is onlife. At the beginning of his book, Floridi refers to a project he managed, a think tank under the name of The Onlife Initiative, established by the European Commission in 2012. The aim of the project was to study the impact of ICT on digital transformation in EU countries. The project bore fruit in the form of a text entitled *The Onlife Manifesto. Being Human in a Hyperconnected Era,*[5] compiled in 2013 and then published in a book of the same title, edited by Floridi, containing a range of essays addressing the issues encompassed by the notion of onlife (Floridi 2015). And the notion in question means being present in the world, with this presence being both analog and local and digital and *online* at the same time. In such a world, the borders between the two realities become blurred.

In the introduction to the said book, Floridi summarizes the overall impact of ICT in the form of four fundamental transformations being:

(a) the blurring of the distinction between reality and virtuality;
(b) the blurring of the distinction between human, machine and nature;
(c) the reversal from information scarcity to information abundance; and
(d) the shift from the primacy of stand-alone things, properties, and binary relations, to the primacy of interactions, processes and networks (Floridi 2015, 2).

[5] https://ec.europa.eu/digital-single-market/sites/digital-agenda/files/Manifesto.pdf (accessed on: 10/23/2019).

The said conclusion, which may be considered a kind of synthesis, pertains undoubtedly to digital transformation and is an outcome of studies into it, to which we shall return, but also addresses—although not directly—the phenomenon of digitalization. The blurring of the boundaries between things, which become uniform thanks to the process in question, is its essential quality after all.

Floridi puts forward another notion, which describes the said process: re-ontologization (Floridi 2013). This means a change of existence of the world and things at the deepest—metaphysical—layer, meaning the one which is the basis of their being. This implies that things and the world do not simply adapt to new conditions, but—according to Floridi at least—their substance becomes completely different. On the one hand, it is a good view of transformation. On the other, an accurate image of the role played by the phenomenon of digitalization, which makes re-ontologization, in essence, digital transformation itself.

Re-ontologization is not too difficult to understand. Floridi describes it as an increasingly blurred and unclear interpenetration of what we consider to be present *here*—what we find close and well-known: analog, carbon-based, and *offline*, with what we consider to be *there*—foreign, digital, silicon-based, and *online*. As I am writing these words, not even 10 years have passed since Floridi offered the said image. Yet, I tend to feel it is already somewhat naive. And not because I find it overly simple, but because this interpenetration has gone much further, and both of its ingredients merge in a much deeper, more complex, and—ultimately—more fundamental manner. Re-ontologization has become a fact that has lost its uniqueness, becoming a new natural environment. A new world we live in.

Such a conclusion, not at all pointless, challenges many of the analyses of the phenomenon of digitalization described herein in a way that it exposes its superficiality. And it may, in some conditions, lead to mistakes, to inconsistency, or to inconclusivity at best. Of course, these flaws will not be apparent in the case of uncomplicated, undemanding analyses, based on simplified grounds. Like those accompanying the majority of business literature. But they may cause some unexpected effects in certain unexpected situations.

A good example of such a simplification is Euclid's geometry, inverted around the third century BC, remaining in official use until the end of the nineteenth century. The work entitled *The Elements*, where the said geometry is presented and discussed, is the second most often printed book, right after the Bible. Its principles are employed to construct bridges, cars, furniture, glassware, and other three-dimensional creations of human imagination. And yet, this geometry appeared to be too simple at some point to describe the world, which took a dramatic form near the end of the nineteenth century and found a solution in Einstein's theories. Our everyday life calls for a new and much more advanced geometry. One that accepts the existence of phenomena absent from and violating the laws of that of Euclid.[6] But it is necessary to understand the dynamics of the world in extreme scales, which sometimes appears to be necessary.

---

[6] A perfect, not overly complicated, and reliable description of this issue is offered by Penrose (2004).

Digitalization also seems to be a simple technical process, but such a view is merely an outcome of an all too simple research approach. Digital transformation on which digitalization is based may require a completely different scale, a completely different range of conceptual tools to appreciate and describe its weight and consequences.

References

Aristotle. 1933. *Aristotle in 23 volumes, Vols.17, 18*. Trans. Hugh Tredennick. Cambridge, MA/London: Harvard University Press/William Heinemann Ltd.. http://data.perseus.org/citations/urn:cts:greekLit:tlg0086.tlg025.perseus-eng1:7.1032b.

Badiou, Alain. 1988. *L'Être et l'événement*. Paris: Le Seuil.

———. 1990. *Le Nombre et les nombres*. Paris: Le Seuil.

———. 2006. *Being and event*. Trans. Oliver Feltham. 1st ed. London/New York: Continuum International Publishing Group.

———. 2008. *Number and numbers*. 1st ed. Cambridge: Polity.

Bartodziej, Christoph Jan. 2017. *The concept industry 4.0. An empirical analysis of technologies and applications in production logistics*. Wiesbaden: Springer Fachmedien Wiesbaden GmbH.

Bauman, Zygmunt. 1997. *Postmodernity and its discontents*. New York: NYU Press.

Bayer, Greg. 1998. Classification and explanation in Aristotle's theory of definition. *Journal of the History of Philosophy* 36 (4): 487–505. https://doi.org/10.1353/hph.2008.0848.

Bell, Daniel. 1973. *The coming of post-industrial society: A venture in social forecasting*. New York: Basic Books.

Bongiorno, Giorgio, Daniele Rizzo, and Giovanni Vaia, eds. 2018. *CIOs and the digital transformation*. Cham: Springer International Publishing.

Boole, George. 1854. *An investigation of the laws of thought: On which are founded the mathematical theories of logic and probabilities*. London: Walton and Maberly. http://archive.org/details/investigationofl00boolrich.

Brennen, Scott J., and Daniel Kreiss. 2016. Digitalization. In *The international encyclopedia of communication theory and philosophy, 4 volume set*, ed. Klaus Bruhn Jensen, Robert T. Craig, Jefferson Pooley, and Eric W. Rothenbuhler, 556–566. John Wiley & Sons.

Bresnahan, Timothy F., and M. Trajtenberg. 1995. General purpose technologies 'Engines of growth'? *Journal of Econometrics* 65 (1): 83–108.

Castells, Manuel. 1996. *The information age: Economy, society, and culture. Vol 1: The rise of network society. Vol. 2: The power of identity. Vol. 3: End of the millennium*. Oxford: Blackwell.

Ceruzzi, Paul E. 2003. *A history of modern computing*. Cambridge, MA: MIT Press.

Chandler, David, and Christian Fuchs, eds. 2019. *Digital objects, digital subjects: Interdisciplinary perspectives on capitalism, labour and politics in the age of big data*. London: University of Westminster Press.

Davis, Martin. 2000. *The universal computer: The road from Leibniz to Turing*. 1st ed. New York: W. W. Norton & Company.

van Dijk, Jan. 1999. *The network society*. London: SAGE Publications.

Drucker, Peter. 1961. The technological revolution: Notes on the relationship of technology, science, and culture. *Technology and Culture* 2 (4): 342–351. https://doi.org/10.2307/3100889.

Eigner, Martin. 2018. The industrial internet. In *The internet of things: Industrie 4.0 unleashed*, ed. Ulrich Sendler, 133–169. Berlin/Heidelberg: Springer Berlin Heidelberg. https://doi.org/10.1007/978-3-662-54904-9_9.

Flasiński, Mariusz. 2016. *Introduction to artificial intelligence*. Cham: Springer.

Floridi, Luciano. 2010. *The philosophy of information*. Oxford: Oxford University Press.

———. 2013. *The ethics of information*. Oxford/New York: Oxford University Press.

———. 2014. *The fourth revolution: How the Infosphere is reshaping human reality*. Oxford: Oxford University Press.
———, ed. 2015. *The onlife manifesto. Being human in a hyperconnected era*. Cham: Springer.
———. 2019. *The logic of information: A theory of philosophy as conceptual design*. Oxford/New York: Oxford University Press.
Floyd, Thomas L. 2014. *Digital fundamentals*. 11th ed. Boston: Pearson.
Foucault, Michel. 1969. *L'archéologie du savoir*. Paris: Gallimard.
———. 1971. *L'ordre du discours: leçon inaugurale au Collège de France prononcée le 2 décembre 1970*. Paris: Gallimard.
———. 1972. *The archaeology of knowledge and the discourse of language*. New York City: Pantheon Books.
———. 1981. The order of discourse. In *Untying the text: A post-structuralist reader*, ed. Robert Young, 51–78. Boston: Routledge & Kegan Paul.
Franklin, Seb. 2015. *Control: Digitality as cultural logic*. Cambridge, MA: MIT Press.
Frege, Gottlob. 1980. In *Philosophical and mathematical correspondence of Gottlob Frege*, ed. Brian McGuinness . Chicago: University Of Chicago Press.Trans. Hans Kaal
Fuchs, Christian. 2008. *Internet and society*. 1st ed. Abingdon: Routledge.
———. 2012. *Foundations of critical media and information studies*. 1st ed. London/New York: Routledge.
———. 2015. *Culture and economy in the age of social media*. 1st ed. New York: Routledge.
———. 2016. Critical theory. In *The international encyclopedia of political communication*, ed. Gianpietro Mazzoleni, 1–13. Malden/Oxford: John Wiley & Sons. https://doi.org/10.1002/9781118541555.wbiepc001.
Galilei, Galileo, Orazio Grassi, Mario Guiducci, and Johannes Kepler. 1960. *The controversy on the comets of 1618*. Philadelphia: University of Pennsylvania Press.
Gillies, Donald. 1982. *Frege, Dedekind, and Peano on the foundations of arithmetic*. Assen: Van Gorcum & Co.
Harel, David. 1987. *Algorithmics: The Spirit of computing*. Wokingham/Reading: Addison-Wesley.
Harel, David, and Yishai Feldman. 2004. *Algorithmics: The spirit of computing*. 3rd ed. Harlow/New York: Addison-Wesley.
Hilbert, David. 1950. *The foundations of geometry*. Reprint ed. Trans. E. J. Towsend. La Salle: The Open Court Publishing Company.
Hodgkin, Luke Howard. 2005. *A history of mathematics: From Mesopotamia to modernity*. Oxford: OUP Oxford. http://search.ebscohost.com/login.aspx?direct=true&db=nlebk&AN=177015&site=eds-live.
Horkheimer, Max. 2002. *Critical theory: Selected essays*. New York: The Continuum Publishing Company.
Humphreys, Paul. 2004. Scientific knowledge. In *Handbook of epistemology*, ed. Ilkka Niiniluoto, Matti Sintonen, and Jan Woleński, 549–569. Dordrecht: Springer Netherlands. https://doi.org/10.1007/978-1-4020-1986-9_15.
Ifrah, Georges. 1981. *Histoire universelle des chiffres*. Paris: Seghers.
———. 2000. *The universal history of numbers: From prehistory to the invention of the computer*. Trans. David Bello. 1st ed. New York: Wiley.
———. 2001. *The universal history of computing: From the abacus to the quantum computer*. 1st ed. New York: Wiley.
James, William. 1907. Pragmatism's conception of truth. *The Journal of Philosophy, Psychology and Scientific Methods* 4 (6): 141–155. https://doi.org/10.2307/2012189.
Jameson, Fredric. 1992. *Postmodernism, or, the cultural logic of late capitalism*. Durham: Duke University Press.
Kant, Immanuel. 2004. *Metaphysical foundations of natural science*. Trans. Michael Friedman. Cambridge: Cambridge University Press.
Kline, Morris. 1972. *Mathematical thought from ancient to modern times*. 1st ed. New York: Oxford University Press.

———. 1990. *Mathematical thought from ancient to modern times*. Vol. 3. New York: Oxford University Press.

Krajewski, Stanisław. 2003. *Twierdzenie Gödla i jego interpretacje filozoficzne: od mechanicyzmu do postmodernizmu*. Warszawa: Wydawnictwo Instytutu Filozofii i Socjologii PAN.

Lakoff, George, and Mark Johnson. 1980. *Metaphors we live by*. 1st ed. Chicago: University of Chicago Press.

Legner, Christine, Torsten Eymann, Thomas Hess, Christian Matt, Tilo Böhmann, Paul Drews, Alexander Mädche, Nils Urbach, and Frederik Ahlemann. 2017. Digitalization: Opportunity and challenge for the business and information systems engineering community. *Business & Information Systems Engineering* 59 (4): 301–308. https://doi.org/10.1007/s12599-017-0484-2.

Leibniz, Gottfried-Wilhelm. 1740. *Kleinere Philosophische Schriften*. Jena: Mayerischen Buchhandlung.

Lenzen, Wolfgang. 1991. Leibnitz on properties and individuals. In *Language, truth and ontology*, ed. K. Mulligan, 193–204. Springer Science & Business Media.

Lexico.com. 2020. Transform | Definition of Transform by Lexico. 2020. https://www.lexico.com/en/definition/transform.

———. 2022. Digital | Definition of Digital by Lexico. Lexico Dictionaries | English. 2022. https://www.lexico.com/en/definition/digital.

Machlup, Fritz. 1962. *The production and distribution of knowledge in the United States*. Princeton: Princeton University Press.

Maciąg, Rafał. 2012. *Deus ex machina: nowe media i ich projekt poznawczy*. Kraków: Wydawnictwo Uniwersytetu Jagiellońskiego.

———. 2016. *W stronę cywilizacji Internetu: zarządzanie w naukach humanistycznych*. Kraków: Wydawnictwo Uniwersytetu Jagiellońskiego.

Makkreel, Rudolf A. 1973. *Dilthey: Philosopher of the human studies*. Princeton: Princeton University Press.

Masuda, Yoneji. 1981. *The information society as post-industrial society*. Washington, DC: World Future Society.

Mazlish, Bruce. 1995. *The fourth discontinuity: The co-evolution of humans and machines*. New Haven: Yale University Press.

Merriam-Webster.com. 2019a. Definition of DICTION. 2019. https://www.merriam-webster.com/dictionary/diction.

———. 2019b. Definition of DIGITS. 2019. https://www.merriam-webster.com/dictionary/digits.

Mikolov, Tomas, Ilya Sutskever, Kai Chen, Greg Corrado, and Jeffrey Dean. 2013. Distributed representations of words and phrases and their compositionality. *ArXiv* (October). http://arxiv.org/abs/1310.4546.

Müller, Vincent C., ed. 2018. *Philosophy and theory of artificial intelligence 2017*, Studies in applied philosophy, epistemology and rational ethics. Vol. 44. Cham: Springer International Publishing. https://doi.org/10.1007/978-3-319-96448-5.

Murawski, Roman. 2004. Mathematical knowledge. In *Handbook of epistemology*, ed. Ilkka Niiniluoto, Matti Sintonen, and Jan Woleński, 571–606. Dordrecht: Springer Netherlands. https://doi.org/10.1007/978-1-4020-1986-9_16.

———. 2010. *Essays in the philosophy and history of logic and mathematics*. Amsterdam: Editions Rodopi B.V.

Nagel, Ernest, and James R. Newman. 1958. *Gödel's proof*. New York: New York University Press.

Negroponte, Nicholas. 1995. *Being digital*. London: Hodder & Stoughton.

Olleros, F. Xavier, and Majlinda Zhegu. 2016. Digital transformations: An introduction. In *Research handbook on digital transformations*, ed. F. Xavier Olleros and Majlinda Zhegu, 1–20. Cheltenham: Edward Elgar Publishing.

Olszewski, Adam. 2009. *Teza Churcha. Kontekst historyczno-filozoficzny*. Kraków: Universitas.

Papacharissi, Zizi, ed. 2010. *A networked self: Identity, community, and culture on social network sites*. 1st ed. New York: Routledge.

Peckhaus, Volker. 2003. The pragmatism of Hilbert's programme. *Synthese* 137 (1/2): 141–156.

Penrose, Roger. 2004. *The road to reality: A complete guide to the laws of the universe*. London: Jonathan Cape. Random House Group Ltd.

Penrose, Sir Roger. 1999. *The Emperor's new mind: Concerning computers, minds, and the laws of physics*. Oxford: Oxford University Press.

Russell, Stuart J., and Peter Norvig. 2010. *Artificial intelligence: A modern approach*. Upper Saddle River: Pearson Education, Inc.

Schallmo, Daniel R.A., and Christopher A. Williams. 2018. *Digital transformation now! Guiding the successful digitalization of your business model*, Springer briefs in business. Cham: Springer International Publishing.

Scheer, August-Wilhelm. 2017. Theses on Digitalization. In *The Drivers of Digital Transformation: Why There's No Way Around the Cloud*, ed. Abolhassan, Ferri, 33–43. Cham: Springer International Publishing.

Schreckling, Edward, and Christoph Steiger. 2017. Digitalize or drown. In *Shaping the digital enterprise: Trends and use cases in digital innovation and transformation*, ed. Gerhard Oswald and Michael Kleinemeier, 3–27. Cham: Springer International Publishing. https://doi.org/10.1007/978-3-319-40967-2_1.

Schwab, Klaus. 2016. *The fourth industrial revolution*. New York: Crown Business.

Schwab, Klaus, and Nicholas Davies. 2018. *Shaping the fourth industrial revolution*. Geneva: World Economic Fund.

Sendler, Ulrich, ed. 2018. *The internet of things*. Berlin/Heidelberg: Springer Berlin Heidelberg. https://doi.org/10.1007/978-3-662-54904-9.

Snow, C.P. 1962. *The two cultures: And a second look – An expanded version of the two cultures and the scientific revolution*. Cambridge University Press.

Sorensen, Roy. 2005. *A brief history of the paradox: Philosophy and the labyrinths of the mind*. Revised ed. Oxford: Oxford University Press.

Turing, A.M. 1937. On computable numbers, with an application to the Entscheidungsproblem. *Proceedings of the London Mathematical Society* s2-42 (1): 230–265. https://doi.org/10.1112/plms/s2-42.1.230.

Turkle, Sherry. 1995. *Life on the screen: Identity in the age of the internet*. New York: Simon & Shuster Paperbacks.

Varanini, Francesco. 2018. Human being in the digital world: Lessons from the past for future CIOs. In *CIOs and the digital transformation*, ed. Giorgio Bongiorno, Daniele Rizzo, and Giovanni Vaia, 11–45. Cham: Springer International Publishing.

Vogelsang, Michael. 2010. *Digitalization in open economies. Contributions to economics*. Heidelberg: Physica-Verlag HD.

Wachal, Robert. 1971. Humanities and computers: A personal view. *The North American Review* 256 (1): 30–33.

Warwick, Kevin. 2011. *Artificial intelligence: The basics*. 1st ed. New York: Routledge.

Wilder, Raymond L. 1967. *Introduction to the foundations of mathematics: Second edition*. 2nd ed. New York/London/Sydney: John Wiley & Sons, Ltd.

Chapter 2
The Practice of Digital Transformation

The story of digital transformation does not seem to be particularly complicated if we consider the publications that attempt to describe it overtly, starting as early as from the title. In most cases, digital transformation takes on the form of a phenomenon that materializes well in the minds of authors. One can, of course, argue that this form is incomplete, or quite the contrary—that it is illegitimately vast, but this form emerges each time in a cohesive shape. However, if one comes to realize the variety of contexts appearing here, it would seem we are dealing with a quite well-known and familiar situation. The individual analyses of digital transformation, each with an own order but mostly similar to one another in their assumptions, shall be treated as typical isolated approaches of a science based on the old and common ordinary academic cause-and-effect approach. The analytical structures are spread out usually following a well-known effect dynamics.

Google Trends sees a quick growth in the interest in the idea of digital transformation taking place around 2015, which should be considered a symbolic date, one of little scientific value—for at least several reasons. Firstly, it is a relative measure, resulting from standardization and scaling.[1] Secondly, it is based on a sort of poll, one not controlled or supervised in any way. It is actually dispersed, variable, and—eventually—statistically elusive. These and other circumstances make it impossible to regard it as a reliable scientific source of information. It does, however, give us some idea of the popularity of the analyzed idea, and shows that this popularity is growing faster than before, and the number of searches has clearly increased since a certain moment which took place around the said year.

This is also when the number of books addressing the issue of digital transformation starts to grow. There are very many of them. I am writing these words in January 2020: it is easy to find well over fifty English-language publications published from to 2018 to 2020, featuring digital transformation in the title (they are listed in the

[1] https://support.google.com/trends/answer/4365533?hl=en&ref_topic=6248052 (accessed on: 1/2/2020).

R. Maciag, *Knowledge as a Tale*, https://doi.org/10.1007/978-3-031-69820-0_2

Appendix). There are at least two times as many works claiming to be dealing with the subject of digital transformation or digitality. Such a number makes a thorough, in-depth review quite pointless. We are facing an intense surge that is better not to be opposed. A striking feature of most of them is the type of adopted perspective, which is usually about running a business in new, digital conditions. Such a perspective is a manifestation of the manner in which the world is interpreted within the framework of clearly emerging new and popular trends. According to the vast majority of accounts, with some of them to be discussed here, the aspiration to an efficient, effective, and—ultimately—profitable business is the first but not the only one factor here. The selection of the books featured below and discussed in more detail is a statistical sample, accompanied by publications that seem to be particularly interesting.

The definitions of digital transformation also seem to increase in number. Schallmo and Williams offer their own list (Schallmo and Williams 2018, 10). I have collected the following:

- "The digital transformation can be understood as the changes that the digital technology causes or influences in all aspects of human life" (Stolterman and Fors 2004, 689);
- "Digital transformation is a continuous complex undertaking that can substantially shape a company and its operations" (Matt et al. 2015, 341);
- "Digital transformation is a new development in the use of digital artifacts, systems and symbols within and around organizations" (Bounfour 2016, 20);
- "Digital transformation is the changes associated with the application of digital technology in all aspects of human society" (Leodolter 2017, 198);
- "We define digital transformation as a sustainable, company-level transformation via revised or newly created business operations and business models achieved through value-added digitization initiatives, ultimately resulting in improved profitability" (Schallmo and Williams 2018, 4).

The list comes supplemented by a more extensive description offered by Nagy Hanna: "Digital transformation signifies deep structural changes in the economy and society brought about by harnessing the full potential of the ICT revolution, and its by-product, the information revolution, with the aim of building smart economy and society. These changes typically lead to increased use of information and communication resources and technologies; reduced cost and increased the pace of transactions, and interaction throughout the economy; full or partial digitization of information, products, services, and processes; empowerment of individuals, communities, and enterprises with connectivity and knowledge; and increased share of information-based production and knowledge services in society. Transformation, therefore, is a radical improvement of societies' capabilities to get things done – to shape their physical, economic, social and intellectual environments to their own ends." (Hanna 2016, First edition:27).Apart from a pure definition, it includes a conclusion concerning the impact of the considered phenomenon on the social world, and formulates a sort of forecast. It is one of the boldest and most resolute opinions about digital transformation, but similar in its nature to other ones on account of its slightly premature enthusiasm and optimism.

One of the first texts on digital transformation, opening the selection of publications listed herein, is a chapter written by Erik Stolterman and Anne Croon Fors, featured in a book from 2004 (Stolterman and Fors 2004). This is a time of a very early reflection, a reflection far from harbingering the eruption of texts ten years later. Which makes the interpretation of Stolterman and Fors all the more interesting. They see their task in fundamental terms, which is a result of a very serious approach to the issue of digital transformation. The researchers display something they refer to as "research position", designed with a very broad perspective in mind. Their starting point is digital technologies and the impact thereof on all aspects of human life, meaning their approach is rather trivial, deterministic. Yet, they also assume that they are dealing with a certain "greater whole", intense and multi-scale phenomena, something "that is increasingly experienced with, through, and by information technology", leading to an important conclusion: "What we are witnessing is an ongoing radical digital transformation" (Stolterman and Fors 2004, 659). Digital transformation becomes thus much grander, much more significant, and explains the idea of a "greater whole".

The notion that sufficiently encompasses this whole is life. Or—speaking more precisely—"good life", being the final instance, the framework of reference for the reflection made. Such a task involves a strong ethical aspect and shifts the attention towards the human, who appears in the subjective, the existential, and the social perspective alike. The said authors declare to be following a critical approach, which also includes aesthetic matters. The research position they adopt is thus very general, clearly philosophical by default. This kind of selection is the most important message of the text, which itself makes references to an old European humanistic scheme formulated eventually by Kant.

The message, naive from the point of view of business pragmatics, had become one of the statements initiating the reflection on digital transformation. However, within the domain of the subject matter of this transformation viewed in strict terms, the attitude in question did not become dominant, defeated by the practical and purposeful current of managerial thinking. In 2015, Matt et al. summarized business strategies connected with digital transformation in an article addressing the very issue. They base their argument on a number of publications published since 2013, coming to a conclusion that the problem is broad indeed. It is particularly challenging to identify the areas it concerns—and for a simple reason: digital technologies can be applied to any undertaking a company may be faced with:"Digital transformation strategies take on a different perspective and pursue different goals. Coming from a business-centric perspective, these strategies focus on the transformation of products, processes, and organizational aspects owing to new technologies" (Matt et al. 2015, 339). Their interpretation of digital transformation refers to technologies that play a trivial role of a reality-transforming determinant. But the range of these changes calls for a less trivial approach. In the face of the said areas and tasks where the said technologies may cause radical changes, it becomes clear that a simple deterministic perspective is not enough.

Matt et al. propose an analytical structure: Digital Transformation Framework (DTF), based on the idea of four dimensions, which make it possible to describe the development of an enterprise—with a considerable approximation. Appropriate measures taken in different places position such an enterprise within the adopted dimensions, describing its relationship with the said transformation. The latter is not perceived anymore as a set of actions or tasks, but as "a continuous complex undertaking that can substantially shape a company and its operations" (Matt et al. 2015, 341).

The said four dimensions, which shall be treated as consistent and interconnected directions of action, are the following: use of technologies, changes in value creation, structural changes, and financial aspects (Matt et al. 2015, 340). They describe certain cross-organized elements of company management. Financial aspects are a specific center that connects value creation and structural changes. The center remains in a basic relationship with the fourth dimension: technologies. To Matt and his associates, the notion of dimension has a rather pragmatic meaning. Operating a company is a kind of a search for balance between these dimensions, which are actually certain describable or settable challenges. The notion of dimension means a metaphorical field of issues, not a scale of potential values. These dimensions do not form an abstract, four-dimensional space. Matt and his associates' analysis does not follow a similar generalization. The world of their analysis is a flat plane because the structure, the framework that describes it is only a group of relationships between different processes. It merely approximates the image of space, and does not cross the boundary that would make this image an object of imagination or thought, which would then turn it into an abstract model of digital transformation.

The idea of dimensions is elaborated on further in a 2016 book by Ahmed Bounfour (2016). The book closes and summarizes a research program named Information Systems Dynamics (ISD). The program had been carried out since 2010 by CIGREF, a foundation acting as a scientific body to CIGREF-Association (Club Informatique des Grandes Entreprises Françaises), bringing together 130 of the largest European coroporations . These circumstances are, of course, vital, and offer an important adjustment in terms of time. Digital transformation was an object of keen interest much earlier than according to what the findings of Google Trends suggest. It is an occasion to make a certain revision of the historical development of the idea of digital transformation in literature.

The axis of Bounfour's analysis is the aforesaid idea of dimensions, which arises essentially from the way in which digitization is perceived and described. However, the source on which he bases his approach is quite different. Digital transformation naturally echoes the fundamental aspects of the process in question. Bonfour formulates these aspects based on an article featured in a special edition of "Management Information Systems Quarterly", entitled *Digital Business Strategy: Toward a Next Generation of Insights*—published in June 2013. It is the year in which the interest in digital transformation, resulting later in a serious reflection, gains momentum.

Bounfour refers in particular to the text by Bharadwaj et al., which is the introduction to the said edition (Bharadwaj et al. 2013). The notion of digital transformation is completely absent from it. The primary object of reflection is, in turn, a superior phenomenon based on a combination of IT strategies and business strategies, which the authors named *digital business strategy* and defined in working terms as an "organizational strategy formulated and executed by leveraging digital resources to create differential value" (Bharadwaj et al. 2013, 472). The key to this new phenomenon is certainly "digital technologies [which] are fundamentally reshaping traditional business strategy as modular, distributed, crossfunctional, and global business processes that enable work to be carried out across boundaries of time, distance, and function (. . .). Digital technologies are also transforming the structure of social relationships in both the consumer and the enterprise space with social media and social networking" (Bharadwaj et al. 2013, 472).

Such kind of inspiration gives birth immediately to major consequences. First, the nature of the definition and the expressions used therein place it categorically in the domain of hard business management. Second, the selected area of digital business is a phenomenon that will soon become so distinct and complex that it will need a better name, and this name is digital transformation—a notion not that neutral and much more dramatic in expression. Third, Bharadwaj and his associates' text articulates the connection existing between earlier publications dealing with digitality in the economic context but not using this notional invention. The last observation makes it possible to deepen the historical sources of the thought that eventually leads to the emergence of the invention in question, considering the publications mentioned by Bharadwaj et al. They date back to as early as 2001, but not earlier, when they formulate the noticed highly fundamental transformation of business as prompted by digital technologies.

The starting point for the inspiration followed by Bounfour is four themes formulated by Bharadwaj et al., serving to provide a general framework for reflection on digital business. These are: "(1) the scope of digital business strategy, (2) the scale of digital business strategy, (3) the speed of digital business strategy, and (4) the sources of business value creation and capture in digital business strategy" (Bharadwaj et al. 2013, 472). Bounfour interprets them in the spirit of management, which is based mainly on the idea of so-called value added, which simply means a gradual formation of a creative element in a produced good, developed and eventually sold, becoming, in the end, the basis of profit.

Bounfour elaborates on the idea of themes to a great extent. On the one hand, thanks to extensive empirical research, he identifies the most important trends that can be observed in business reality. On the other hand, he uses them as a basis for an abstract synthesis leading to conclusions in the form of a new type or organization, based on a modernized standard—in the broad sense of the term—of the manufacturing process, viewed in a new setting, being the digital reality and its medium—data. The concept is offered against the background of a historical analysis, based on a very elementary understanding of the main object of control and management: land in the feudal system or labor in the industrial system.

In the digital reality of today, this role is played by a virtual resource of broadly defined "links" in the network of which a company exists and operates. The inspiration is surely the Internet, but the network is also composed of other relationships, which make up a complex and dynamic structure. The resource of links forces new strategies and new types of organizations, which means a transformation of the entire economic system—and the social system as well. Bounfour calls the emerging phenomenon "accelution", underlining the role of acceleration, which involves changeability (liquidity), or even the liquidation of the boundaries of time and space. "Accelution" is "the emergent mode of production, founded on digital, in which the key resource is fundamentally immaterial: it consists of increasingly accelerating links, which the economic operators seek to control" (Bounfour 2016, 79).

Bounfour borrows the idea of liquidity from Bauman (2000), which means he intends to—and does—move within a set of notions going far beyond the area of management. He seems to understand that what is at stake in the new, digital world is not only economic performance, and that the discussion takes place on a much more general, cultural level. The presence of such a level of reflection stems from the observations made, summarized in the book, which are successfully reconstructed in an extensive and neatly arranged taxonomy of problems to which Bounfour refers to as—according to the nature of the new paradigm based on mobility and liquidity—trends.

Bounfour defines twenty five different trends of development and change in contemporary enterprises, grouped into seven more general categories. This contributes to the erection of a large-scale interpretation structure, which remains valid and relevant in its vast majority, even though it's purely pragmatic—or even utilitarian—approach is surely highly biased and one-sided, which does not mean limited. The description of this structure covers many pages of Bounfour's work, acting as an extended basis for synthetic conclusions.

Here are the seven categories and twenty five different trends of development and change according to Bounfour:

Emerging Business Models

1. Business models are undergoing a major transformation, without any fundamental difference appearing between purely digital models and traditional models.
2. The expansion of value creation spaces is a major trend driven by digital.
3. The "customer experience", mediated by digital platforms, is an essential dimension of emerging digital uses.
4. "Pure" digital business lines (software and services) appear to be quicker to develop adaptive offerings than more "traditional" business lines.
5. New business models call for new managerial skills.

Work, Coordination and Digital Uses

6. In enterprise 2.0, hierarchy and horizontality (community) are not naturally opposed: they are two distinct but complementary modes of coordination.
7. Digital technology is relatively neutral with regard to collaborative uses; uses are also (above all?) driven by organizational specificities. Individuals adapt to the nature of technology by improvisation, especially where the technology is flexible.

8. Mobility: the revolution is already visible, but its transformational impacts are still taking shape.
9. The behaviors of the new generations (Y and C[2]) reflect new digital uses... but they also derive from earlier managerial practices.
10. The impact of digital uses on the question of control needs to be approached indirectly, taking due account of the nature of tasks.

Internal Innovation Practices

11. Innovation is a complex process, in which top-down approaches intermesh with bottom-up approaches.
12. Internal collaboration on innovation can be far from straightforward, especially in contexts dominated by individual incentives.
13. Digital develops a view of the enterprise as a knowledge market, but it also requires the development of approaches centered on more organic relationships.
14. The development of innovative digital products and services in a geographically distributed context calls for continuous and repeated management of alignment tensions (between teams, disciplines and functions).

Open (External) Innovation Practices

15. Open innovation practices contribute to performance. They are heavily reliant on—and greatly facilitated by—companies' digital infrastructures.
16. Joint collaborative platforms emerge in competitive industries by means of ad hoc learning processes.
17. Online collaboration to develop knowledge can be organized even between people with no previous contact.
18. Spot markets, crowdsourcing, communities and hierarchy are complementary spaces, and their overall governance remains to be defined.

Enterprise Space and Knowledge Flows

19. In digital space-time, remote collaboration calls for skills-sharing to be redesigned. Ad hoc (agile) design methods make this possible.
20. In the digital economy, IT entrepreneurship requires the creation of ad hoc spaces of freedom, especially by large companies.

The Social and Ethical Dimensions of Use

21. For work-related uses, the ethical issues are still emergent. Users are aware of the importance of the topic, but "double-loop" learning is not yet fully in place.
22. While mobile uses and ubiquitous connectivity may pose ethical problems for public users, this is far from being a universal issue, especially in Asia.

---

[2]According to Howe and Strauss' theory (Howe and Strauss 1991), Generation Y (so-called Millennials) includes people born in the years 1981–1996, and Generation C—defined less formally and in different ways—is the former's approximate successor.

23. The hypothesis of the end of privacy, by a general over-exposure of individuals, is not confirmed. Individuals adjust their exposure behavior to the policies of operators like Facebook, and indeed to the structure of their own social networks.

Data, Intellectual Property, and the Specificity of Digital

24. Digital induces tensions between regulation and freedom, and between privacy and the freedom to do business. Copyright, in particular, is endangered by the development of open-source and open-access practices.
25. Digital is a generative machine that produces spontaneous, unexpected innovations, often contributed by external developers.
26. Network abundance multiplies the tensions involved in organizing the enterprise (fly-by-wire vs. decision support; security vs. privacy; ownership vs. profitability; public goods vs. private goods) (Bounfour 2016).

In their conception, Bounfour's trends have a holistic nature; they include both the inside and the outside of an organization, consider the different aspects of its presence, but do not lose their central, business-oriented perspective. The introduction of such notions as time or space seems to direct Bounfour's imagination towards a more advanced interpretation of the modus governing the existence of an organization on the one hand, with this organization becoming an increasingly abstract being on the other. This kind of thinking is crowned by an image in which an organization becomes suspended in an abstract sphere being the essence of digitality: the ecosystem of data, characterized by five dimensions setting the directions of digital transformation. These are: "(1) the expansion and plurality of spaces; (2) the articulation between transactional links and organic links; (3) spatio-temporal structure; (4) organizational plasticity, and finally the central element of the analysis (5) the acceleration of links" (Bounfour 2016, 70).

It is a very interesting vision, especially because of its very general and abstract nature. Both the sphere and the organization placed within it break away almost completely from the material, realistic experiential attitude. Instead, they become theoretical beings, characterized respectively by: (1) the modus of their existence: hard to locate, dispersed, (2) the relationship with the surrounding environment: based on homogeneous and unlimited relations, (3) the ontology: aware of its physical universality, (4) the manner in which it functions: changeability, and (5) its nature: dynamic and networked. A relevant illustration depicts the positions of these dimensions, which appear to be, however, only metaphorical, taking the form of a flat plane. These dimensions appear there only as fragments, which greatly impairs Bounfour's abstract concept, although one of the most interesting and most advanced in the process of description of digital transformation.

Bounfour's in-depth analysis is a testimony to a nomological approach which is analytical and abstract at the level of conclusions; an approach that aims at generalizations and, basically, presents them as such. But this is not the way the vast majority of the reflection on digital transformation had followed. It did not defend itself against pragmatism closed in the subject of analysis, which, like an explosive, waited to turn into a grand outburst of publications on the materialization of digital transformation treated as a business-oriented, urgent process of company reconstruction. This great tide, rising since 2018, includes many ordinary textbooks, but there were also interesting currents or singular examples of a mature and disciplined consideration.

Before we move on to one of the more important issues, inspired by and centered on expert knowledge of technology, let us reach for an example of another variation of reflection on digital transformation. The book by Schallmo and Williams, which we have already discussed at some length before, is unquestionably also a kind of textbook. It was conceived as a guidebook, making use of examples, but it is also based on a professional analysis, the core notion of which is a business model (Schallmo and Williams 2018). It is not a neutral notion. It immediately introduces this unbearably practical type of thinking, which seems to be quenching any other reflection aimed rather towards softening the applied notions and subjecting them to an ambiguous and vague exegesis, hoping for a bright and unlimited world of associations and meanings related thereto. No. It is an idea of technocrats, those deprived of poetic, philosophical imagination. We have at least two reasons NOT to give any attention to Schallmo and Williams' book. But they are counterbalanced and eventually defeated by the authors' discipline, based on an adopted notion, one with a well-defined designatum.

Schallmo and Williams define their basic notion—business model—as the fundamental logic of a company, one "which describes what benefits are provided to customers and partners" (Schallmo and Williams 2018, 11). This is a very clear definition from the domain of management, making the entire book follow a similar path style-wise. But to complete this mission, the authors provide quite an extensive context with an assumption of a fundamental nature. They begin with an interpretation of the term "digitality", which we have already addressed, also elaborating on the alternation of digitization versus digitalization. Next, they refer to existing literature dealing with the subject of digital transformation, quoting selected definitions thereof. This is when they propose their own definition, based on an explanation of their core notion of business model, which makes the overall tone related to business and managerial approach become more stable and specific. This way, a certain synthesis of the digital transformation they describe becomes possible, turning into a well-defined interpretation framework. This is accompanied by three empirical examples, which act as a starting point to the authors' own—and very broad—proposition of a so-called "road map": a set of steps to take to implement digital transformation successfully.

The latter phenomenon ceases to play a part of a vague metaphor, but takes a very pragmatic, even literal shape of a specific process aimed to simply transform an 'analog' company into a digital one. The stages of these processes, named not necessarily exactly literally, are: Digital Reality, Digital Ambition, Digital Potential, Digital Fit, and Digital Implementation. Their descriptive and slightly emphatic nature hides quite simple measures, such as environmental analysis, designing goals, developing digital tools, the evaluation of the potential of these tools, and the utilization of these tools in practice. They are most surely interesting to practitioners, explained and illustrated in sufficient detail. The key to this culminating point, however, is the prior analysis of the phenomenon of digital transformation.

Schallmo and Williams finish the introductory, analytical and theoretical part of their book with a rather succinct conclusion, which is actually an extended version of their definition of digital transformation. It consists of five conceptual factors, formulated as essential questions to make it easier to understand them. The answers to these questions, arranged in the form of short lists, make up the content of the said definition. The structure is clear and internally consistent. Among the factors such as: procedure, which answers the question of how digital transformation occurs and exists, degree—how intense the transformation is, reference unit—for whom the transformation is conducted, or object—what is transformed, there is one more—primary—factor: objective dimensions (Schallmo and Williams 2018, 13).

This factor is actually essential because it sets the further path of analysis. Moreover, it hides an actually very simple thing: the circumstances of occurrence of (the need for) digital transformation, but the authors refer to it as more general and abstract “dimensions”. It is not the first time in which we come across this notion in the context of digital transformation, and it is not the first time it appears to have exactly such a nature. It is also the only factor from among those proposed by the authors to be of a two-aspect nature. Apart from specific materializations, some real, business facts, such as quicker manufacturing, reduced costs, automation, networking, change in the qualitative relations concerning products or customer relationships, etc., there is a level in which they are all grouped together in four areas referred to as dimensions. These dimensions are: time, finance, space, and quality. Their presence proves that digital transformation as a certain phenomenon and as a designatum of such a notion does not fit into the outline of simple inclusion. In other words, it cannot be expressed as a list of features, circumstances, etc., but explodes to the level of abstract generalization, which appears to be once again its immanent quality. Even if the intentions of its analysts are as pragmatic as in the case of Schallmo and Williams’ book. And it explains their slightly exaggerated emphasis of their notional constructs.

By referring to Schallmo and Williams’ book, we abandon the area of management in its literal sense. It is an area very rich in numerous interpretations, but also one quite monotonous at the same time. Yet, this does not mean that it does not enjoy interest, common, for instance, among profit-oriented undertakings. Its pragmatism, highly trivial apart from a few exceptions, is so utilitarian and opportunistic that it, in fact, completely ignores the important existence and effects of digital transformation. This conclusion does not concern, however, the theme based on a question about the technique, gaining an insight into its actual foundation through this direct and honest simplicity. One should not expect, however, a Heideggerian perception or grandeur in terms of interpretation. The reflection is usually careful and detailed with one exception: when it is formulated in the context of schemes of great scale and range, encompassing areas such as political or civilizational systems. They are referred to as: Industry 4.0, cyber-physical systems, or the 4th Revolution.

Before these schemes become described here in more detail, it seems reasonable to discuss three books published in 2016 and 2017, a moment before the said outburst of publications. All three of them try to find a synthetic and each time different key to the description of digital transformation emerging from the domain of technology. These are evidence of ambition and failure at the same time.

The authors of the collection of texts published collectively entitled *Digital Transformation: Information System Governance* (Leignel et al. 2016)—this is the first of the said books—date back as far as to the 1960s, retracing the beginnings of digital transformation. The arrival of big computers with considerable computing power back then helped automatize a number of business processes, such as accounting or stocktaking—as evidenced by the example of France in particular. Two further important steps in this domain occurred in the 1980s, which brought about an increasing popularity of desktop PCs and electronic communication, and in the 1990s, which saw the emergence of special customer relationship management (CRM) and Business Intelligence (BI) software. This brief retrospection shows us the very specialized way of treating the change connected with digitalization. Information technology systems emerge in the form of real computing machines, equipped with appropriate software, thus irreversibly entering the various domains of business. IT departments gain a new—and much broader—significance in shaping business; they change from customer service departments to managing partners, and the relationship eventually becomes completely reversed: "The digital transformation can succeed only if it is supported by an 'IT-oriented' corporate culture" (Leignel et al. 2016, xx).

For Leignel et al., the notion of IT systems is a synonym and generalization of the presence of digital technologies of digital transformation. Maedche et al., the authors of the second book, propose another line of interpretation, based on the idea of Design Science Research (Maedche et al. 2017). Their book is the fruit borne out of a 2017 conference, with the main theme being: "Designing the Digital Transformation". The core idea was "to solve design problems through the innovative use of information technology and applications" (Maedche et al. 2017, V). It is surely a reference to the idea proposed by R. Buckminster Fuller as early as in 1957 (Fuller 1957), formulated in the field of architecture, based on a generalization of the process of design, one which eventually departs from its specific area of materialization. The idea also refers to information technology systems, which are also based (metaphorically) on a structure called architecture since they combine many components of different nature into a functional whole and have to fulfill various expectations and serve a number of purposes. They are thus most definitely an undertaking calling for a general approach. In the case of the set created by Maedche et al., *design* acts as a cognitive category organizing the problem of digital transformation understood as a technological undertaking, but one too complex to consider it only as a loose collection of technical solutions.

An opposite strategy is advocated for by the author of the third book, Ferri Abolhassan, who picks one technology, but of great—according to him—importance. The different contexts of application of this technology are then covered in more detail by him and his fellow researchers. It becomes a technical foundation—or, looking from a different perspective, the apex of digital transformation. Abolhassan explains it as follows: "Anything that can be digitized will be digitized, and anything that can be networked will be networked. This applies to people, machines, and products alike. Software is increasingly becoming a decisive production factor, because all of these networked machines must be controlled and all the digital data must be stored, processed, and meaningfully analyzed. So the successful companies of today also have to be – or become – "software companies." And software lives in the cloud" (Abolhassan 2017, v).

This is a clear interpretation of the problem, based on obvious argumentation, leading to a necessary solution: a "cloud", meaning a contemporary, highly multifunctional technology, which involves transferring almost the entire IT-related part of business operations to rented, remote machines available online (on the Internet).

The technology itself is quite old. In the late 1990s, Amazon offered a service we would refer to today as "cloud-based". In 2012, Williams described it in the following way: "Cloud computing is a model for enabling convenient, on-demand network access to a shared pool of configurable computing resources (e.g., networks, servers, storage, applications, and services) that can be rapidly provisioned and released with minimal management effort or service provider interaction" (Williams 2012, 4).

Abolhassan makes it the core and the axis of digital transformation, giving rise to some concern through such a particular approach. The following question appears: is the world of digital transformation really so trivial? Of course, it would be hard to imagine it without networks, and a network is something that surely opens the door to further opportunities, but it seems that it is exactly the synergistic effect of the diversity of inventions, applications, ideas, processes, and phenomena that contributes to the phenomenon of digital transformation. Also when it is looked at from the point of view of technology.

To close this monotonous part, I would like to refer to a publication that does remain within the framework of the above argumentation, but breaks away from the banality of economic analyses of digital transformation, those concerning management and technology alike. It is a book by Werner Leodolter (2017). It refers to psychoanalysis and is based on a metaphorical construct being organizational unawareness, remaining in a dynamic relationship with the level of organizational awareness. This general pattern of understanding is a response to the development of artificial intelligence and deep learning, leading to the emergence of hybrid intelligence solutions based on man-machine collaboration. The perspective offers a holistic vision of organizations in a situation of expansion of all and any artificial and autonomous systems based on knowledge, but also makes it possible for human to maintain control and leadership.

Leodolter draws the reader's attention to human resources and competencies, which will enable humans to build an effective counterpoise and support machines successfully—including: "experience based on perception with all senses, common sense, metacognition, morals, generalisation, imagination, dreaming, creativity, abstraction, metaphorism, compassion, empathy, emotion, solving dilemmas etc." (Leodolter 2017, 251). Such a point of view opens the door to a plane of standardization of natural and artificial cognitive systems by providing a common platform of such a type at the organizational level, one understood as a metaphorical subject, equipped with mental and psychological qualities. Digital transformation is regarded as a process functioning at such a specific level of abstraction. Such a perspective imparts a deep meaning upon it, similar to the one proposed herein, and at the same time enables us to make rough, or even fundamental, observations, like those concerning humankind. Leodolter claims that: "The digital transformation is said to be an evolutionary step for mankind sometimes resulting in disruption – for example in entirely new products, services and business models –, but definitely resulting in paradigm shifts in society. These shifts result from our exponentially accelerating ability to create and process information, to communicate and to share sentiment, ideology and opinion. This is changing the perception and construction of reality and thus affecting and changing the roles of individuals, technology and organizations in the society" (Leodolter 2017, 249).

Grand Schemes: Industry 4.0, Cyber-Physical Systems, 4th Revolution

The following year, 2018, is when publications viewing digital transformation in a much broader perspective start to appear. They are based on phenomena of a different kind, such as Industry 4.0, cyber-physical systems, or the 4th Revolution. Although in practice they are interconnected with one another and bloom in additional ideas and inventions, such as the Internet of Things, they remain a description of certain general and explainable schemes. They also refer to earlier concepts, organizing the processes of digital transformation as if anew.

This is the mechanism on which the book entitled *Industry 4.0: Managing The Digital Transformation* (Ustundag and Cevikcan 2018) is based. In the book, digital transformation is no longer treated as an idea that desperately needs a definition, a notion signifying some vague phenomenon, but a tacitly accepted, obvious technical materialization of a grand government scheme. The idea of Industry 4.0 "was first declared by German government during Hannover Fair in 2011 as the beginning of the 4th Industrial Revolution" (Salkin et al. 2018, 4). The Hannover Fair (in German: Hannover Messe)[3] is one of the biggest annual international fair events, showcasing itself as a major platform for industrial transformation. The event is important especially for business, but its political context is vital as well: it has been run since 1947 as an institutionalized form of representation of the development of industry and economy as a certain whole. In 2019, the fair addressed digital transformation directly, which made the theme become an inseparable part of this kind of subject.

Since 2006, the German government has aimed to establish a solid foundation for the coordination of Germany's research and innovation-oriented initiatives. In

[3] https://www.hannovermesse.de/en/ (accessed on: 1/1/2020).

2011, a research group named Die Forschungsunion Wirtschaft—Wissenschaft,[4] an advisory body to the government, announced the initiation of a relevant project referred to as a government "strategic initiative". The Group offered also a preliminary definition of the proposed initiative[5]: "(...) the fourth industrial revolution, a new level of organization and control of whole value chains over the entire lifecycle of products This cycle includes the fulfillment of individualized customer requirements and extents itself from idea, real order, development, and manufacturing, delivery to the customer and the recycling process with the involved services. The basis for the development is formed by the availability of all necessary information in real-time through interconnection of all instances, which are involved in value creation as well as through the ability to derive the best possible value stream based on the resulting data. Through the connection of people, objects and systems, dynamic, real-time optimized, self-organizing, cross-company value networks will evolve, which can be optimized based on different criteria such as costs, availability and resource efficiency" (Bartodziej 2017, 33).

The notion of "the 4th Industrial Revolution" used in the above definition made the said initiative a kind of a political manifesto, although the idea was not yet so clear and obvious at the time. Bartodziej offers its early definition by Jacobi and Landherr, dating back to 2011, whose aim is digitalization and which has a more general nature—a social change transforming the industrial society into a postindustrial society, one based on services, knowledge, and information (Bartodziej 2017, 33).

The idea of Industry 4.0 has matured and become sharper over time. When it was announced, it focused more on the aspect of management and referred directly to the process of operating a business, although it did formulate a clear technology-inspired demand, based on the idea of a network that makes it possible to coordinate business processes on a great scale. Today, its technological aspects have become prevalent. Salkin et al. mention eight advanced technologies lying at the heart of Industry 4.0: "adaptive robotics, data analytics and artificial intelligence (big data analytics), simulation, embedded systems, communication and networking such as Industrial Internet, cloud systems, additive manufacturing and virtualization technologies" (Salkin et al. 2018, 5). The specialist terminology used in the description also defines the potential professional target audience, although it is underlaid by easily understandable technologies such as universal robots adapting to a given operating environment and a given set task (adaptive robotics), computer control over virtually any technical appliance, including household appliances (embedded systems), or the so-called manufacturing of real objects by "printing" and the related use of digital models instead of their physical prototypes in every task in which such a substitution is possible (3D printing and virtualization).

[4] http://www.forschungsunion.de/ (accessed on: 1/1/2020).

[5] The full text of the report, drawn up three years later, can be found at: https://archive.org/stream/FinalReportRecommendationOnStrategicInitiativeIndustrie4.0/Final%20Report_%20RecommendatIoN%20oN%20stratEGIc%20initiativE%20INDUSTRIE_4.0_DJVu.txt (accessed on: 1/1/2020).

In his book published in 2016, Bartodziej equates both coexisting notions and concepts of Industry 4.0 and the 4th Industrial Revolution with one another, and refers to two underlying technologies behind both of them: cyber-physical systems and the Internet of Things and Services (Bartodziej 2017, 2). These technologies are interconnected. In a book on the fundamentals of cyber-physical systems (Möller 2016), its author Diethmar Möller, presents the history of their development, their evolution since the times of the said embedded computing systems, proceeding through the further degrees of complexity resulting from the linkage of these systems into a network (Möller 2016, 85). Such an innovation, enabling the creation of a universal system of interconnected computers, monitoring and controlling a great number of real-world objects, leads to the logic of a structure founded consciously and intentionally on the principle in question: to cyber-physical systems. Their even more advanced form is the Internet of Everything, encompassing data and services, which means both a passive circulation and source of data and an active tool to satisfy the needs of consumers and companies, including monitoring and diagnosing the health of individuals, or market analyses of companies and of the areas of their greatest, core activity—such as promotion.

The said complexity becomes an outcome of a potential rapid outburst of innovations: a combination of technological solutions and processes of their use, similar to the success of the poor old man going by the name of the Internet, who has been successful since his resurrection in the twenty-first century, after the sudden burst of the speculation bubble on the new tech market in 2000. The mechanism that became the commonly quoted reason behind this success, although it is a great simplification, was the transformation of the unidirectional nature of communication between the seller and the customer. The latter emerged in their uniqueness and ceased to hide in the crowd, which became a group of sharp and detailed portraits of the great majority of us. And these portraits ceased to be dumb, silent. Most of the said communication has actually become reversed: now, it is organizations who listen carefully to consumers and engage them in their operations, and the ideas and technologies supporting these processes are becoming increasingly advanced. An example of the latter is the application of sophisticated tools of social network analysis to study data provided by the so-called social network sites (SNS).

The ever-intense development of the "old" Internet in Möller's model of evolution is transforming: next to people, the said portrait starts featuring everything else—on one condition, of course: it needs to be 'digitalizable'. This is to mean not only the obvious, trivial level of digitization, i.e. digital signals controlling and transmitting data, like in a washing machine, which is not only a mechanical device but a specialized computer which controls the process of washing using valves and dispensers based on various data such as the temperature or the level of water. In order for such a washing machine to be able to work, we need a model of its operation—basically, the very sense of its existence. A virtual model, expressed in the form of an algorithm, a fragment of reality describing its existence in the part of essence in terms of its purpose: the not-so-simple process of washing, which is, after all, a complex sequence of subsequent chemical reactions and physical processes. We could substitute the washing machine with another fragment of reality, such as the processes of reading and understanding, to realize the impact and reach of this model. And this is the area dealt with by neural networks, developing at a great rate before our eyes.

But there needs to be another concept, one simple and unavoidable, to accompany digitization for the described evolution to take place at all: networking. At first, the process seems to be rather banal. The media love telling stories of interconnected fridges and washing machines. But the actual reach and potential of the new network quickly appeared to be much greater; the network gives rise to a potentially unlimited and highly detailed circulation of data. This can be an attractive source of data for a great number of stakeholders, who will represent not only the world of business, determined to improve their products and test their performance on an ongoing basis, monitor the ways in which they are used, their power consumption, or their environmental impact. This data will also make it possible for other, more advanced sources of information to appear. The very same stakeholders will make use of them, of course, to make their outreach campaigns more effective, creating sales and use models of unprecedented precision, resulting from watching their product in action in reality. Such information is, however, very interesting to a much broader group of stakeholders, including centers committed to creating more general social models—useful to a vast number of commercial and government organizations, as well as illegal (usually criminal) organizations, engaged in theft or blackmail, or even more dangerous ones—such as terrorist organizations.

But the really ubiquitous computer systems written about years ago, not only and not mostly because of their technical properties, such as small dimensions, advanced complexity, or low power consumption, meaning not as a result of the simplest level of digitization, contribute to refined grand-scale structures—such as cyber-physical systems. This is determined by their conceptual advancement positioned on the basic and definitional level of abstraction, which is the interaction with the real world. Niklaus Wirth, a classic of the algorithm theory, wrote as follows in his textbook from 1976: "the information that is available to the computer consists of a selected set of data about the real world, namely, the set which is considered relevant to the problem at hand, that set from which it is believed that the desired results can be derived. (. . .). The data represent an abstraction of reality in the sense that certain properties and characteristics of the real objects are ignored because they are peripheral and irrelevant to the particular problem. An abstraction is thereby also a simplification of facts" (Wirth 1976, 1).

The encounter between the computer world and the real world is therefore a challenge and the key to the construction of the former, meaning to the creation of a world of digital, virtual objects, ones different from and, in principle, incomplete with respect to their real-life counterparts. Both of these worlds are also filled with objects for which the boundary between them is impassable; they can exist only on one side, at least for now, because there is an ongoing technological battle fought to make their number increasingly smaller and to make the worlds merge to a greater degree. A lot was written on the subject in the 1990s, with the main inspiration being the phenomena of the cyberworld.

Cyber-physical systems are based on a more pragmatic assumption; without negating the dissimilarity between the world of computers and the real world, they fix themselves consciously on the border between them, exploring both spheres in the part that can be made common. Möller offers a modest definition of the nature of their emergence: "The integration of embedded computing systems – they represent physical systems where the computer is completely encapsulated by the device it controls and interacting with physical processes – with networked computing has led to the emergence of a new generation of engineered systems, the so-called cyber-physical systems (CPSs). These systems use computations and communication deeply embedded in and interacting with physical processes by adding new capabilities to physical systems" (Möller 2016, 81).[6]

There is no doubt that the main, direct, simplest reason behind the effort to unite the worlds in question is the number of shared spaces, the almost tiresome universality thereof, which has already managed to turn into indispensability, irreversibility, irreplaceability, resulting from the opportunities that arise. This is an effect of IT professionals' greedy look on the surrounding world, underlain with a question: what else is there to digitalize? The next reason was provided by the construction of networks which by connecting individual artifacts moved to a higher level of abstraction, where they emerged as mass, large-scale, general processes (all washing machines of the world, unite!). These processes are not only visible, but controllable exactly by means of this network. This way, the vision of a world being a system of controlled large-scale processes appears. A great advantage of this system is the principle according to which the control in question is exercised at the lowest level of individual components, and the tasks carried out concern the uppermost levels, acting on a great scale: cities, states, and higher.

There is therefore no doubt that technologies such as cyber-physical systems and the Internet of Everything need to interfere with processes of a different, more general nature. Speaking more specifically—with social and cultural processes. They thus become a political and ideological issue. Speaking the language of these systems, this is a phenomenon referred to as "social embedding" (Böhle and Huchler 2016). The social context—in the very broad sense of the expression—is already featured in the first chapter of the quoted book (Böhle and Huchler 2016). Its authors start from a claim that the term "cyber-physical systems" appeared in the United States in 2006, initially encompassing issues involved strictly technically in the overcoming of the barriers of the physical world and the cybernetic world. Ten years later, as they argue, it is impossible to find other systems in a reality so extremely subordinate to the logic of digitalization (Torngren et al. 2016, 4). This is a direct argument speaking in favor of conducting their analysis in a much broader context.

Alur notices that some forms of cyber-physical systems have been present in the industry since the 1980s. But it was only the development of technology in the twenty-first century that led to the emergence of an academic discipline named

[6] A clear and in-depth conceptual map of cyber-physical systems can be found on the website of the Ptolemy Project run by the University of California in Berkeley, https://ptolemy.berkeley.edu/projects/cps (accessed on: 1/2/2020).

'cyber-physical systems', acknowledged almost immediately as a field of fundamental significance from the point of view of both industry and governments alike (Alur 2015, 2). These systems introduce new rules concerning the creation of hardware and software solutions, different from those in place within the framework of the classical computability theory, acting as a response to the challenge of collaboration with the physical world, which is no longer suited to the requirements of the computer, but present in its entire richness and unpredictability. These rules are as follows: reactive computation (permanent interaction with the environment), concurrency (opposite to sequential computation in which the Turing machine is a canonical model), feedback control of the physical world (feedback interaction with the physical world), real-time computation (real-time performance), safety-critical applications ("safety of the system has the higher priority over other design objectives") (Alur 2015, 2 passim).

To model the real complexity of physical systems, Torngren et al. propose a structure referred to in short as SPIT (Social, Process, Information, Technology) (Torngren et al. 2016). It encompasses four levels of functioning of these systems, starting from the lowest-tier—technical layer, including all of the essential hardware and infrastructure, through the layer of information, meaning the content of all signals and communication exchanged, the layer of processes which enable the creation of any type of value for users, ending with the social layer. The latter is described as "the organizational and wider context of use of CPS including ethical positions (perhaps regarding autonomy), and regulatory and legal frameworks" (Torngren et al. 2016, 6). It contains the most general level of reflection, meaning one that considers the presence of these systems in the social world.

The aspect is approached from a much broader perspective by Gaisberger and Broy, who argue straight that: "social systems face far-reaching changes as a result of the pervasiveness of Cyber-Physical Systems and the profound impact that they will have on social interactions, processes and structures thanks to their extensive context awareness capabilities (using sensors) and the fact that control of their behaviour can be fully or semi-autonomous (using actuators), as well as being globally networked and integrated into everyday processes" (Geisberger and Broy 2015, 107).They perceive social context as a set of parallel systems that call for an analysis being something much more than just an elongation of the physical aspect of cyber-physical systems, forming complex socio-technical systems as a result. They also abandon the narrow interpretation of the human as a technical element of these systems. This reveals a social structure of a diverse and unpredictable nature, determined by local ethical, cultural, and other factors, and the human is depicted as a being driven by emotions, behaving in an uncharacteristic way, referring to various sorts of values or dependent on communication (Geisberger and Broy 2015, 110). These socio-technical systems are facing a great challenge. This challenge is to identify and provide appropriate services to all actors engaged in their functioning, starting from the most basic level—the human-machine interaction. What makes matters additionally difficult is the fact that cyber-physical systems may also act as certain specific agents of different social actors (e.g. individuals, groups, organizations), operating autonomously and making independent decisions.

The social context of cyber-physical systems is inadvertently aiming at this fragile crux, which is the presence of an individual, faced alone with a highly efficient and sophisticated machine, representing a latent, complex, and powerful system. This quite dramatic situation, rightly picked up and fed on by popular culture, is very real, though. It is a result of such practical circumstances as the ubiquity of the technology, its interactive nature, the lack of transparency of networked technology, and the need for the user to make immediate decisions based on incomplete knowledge, in a hurry, forced by technological processes (Geisberger and Broy 2015, 117).

The relationships between humans and machines cease to be about full control in the hands of the former, but start to involve a common, hybrid or dispersed control, meaning switching from interaction to a collaboration of sorts. This collaboration can take the form of three main modi of mutual positioning:

- The CPS forms a stand-alone system or a framework for human living and action, following the pattern of a functional division of labor. (An example is an autonomous self-regulating production system or an "intelligent house");
- The CPS flexibly responds to human needs and human behavior, following either the pattern of an assistance system or the pattern of customer-oriented supply and provisioning. (Examples are augmented-reality tools or the smart refrigerator);
- The CPS cooperates directly with human actors following the pattern of cooperation in teams at work or customer-orientated interaction in personal services. (Examples are robots in working groups or robots used in healthcare: "carebots") (Böhle and Huchler 2016, 116).

In the context of the described interactions, the act of communication becomes essential as an act that determines the occurrence of mutuality in the human-machine relationship, which involves a mutual exchange of information on the basis of the knowledge possessed by both parties. This knowledge appears to a controversial notion if we recall it's still valid definition dating back to the times of Plato, described in the chapter on the general foundations and sources of reasoning. Knowledge is not just the domain of the human in the context of advanced technology materialized as cyber-physical systems, and occurs in a specific way thanks to communication.

Böhle and Huchler argue for a description of communication on the basis of three main forms thereof: verbal-symbolic communication, action-related communication, and communication through anticipation (Böhle and Huchler 2016, 117 passim). The first is quite obvious: it is language and other symbolic forms. The second, more challenging, is based on meanings assigned to particular actions. An example is the complex behavior of a player on the field, which is an expression of the tactics they have decided to adopt. The third is the least friendly of all, and involves an expectation of signs of behavior known in advance. It is fluid in nature because, in fact, there is no physical process of communication, but a piece of information arising from certain specific circumstances at a given time and place may appear. All three forms of communication may occur between a human and a machine, serving as a mechanism to transfer knowledge.

The growing complexity of cyber-physical systems combined with their increasingly closer relationships with humans will most likely force more refined and unexpected forms of communication. They will contribute to a mutual complementarity and collaboration between technology and mankind, balancing their impact, abandoning the naive, early idea of machines imitating human qualities. The process will take place on a deep level. Böhle and Huchler compare it to the process of structural coupling, whose idea was first proposed by Maturana and Varela (1992, 75). It is based on recurrent interactions between systems, which leads to a mutual conformity and concordance. In such conditions, of course, the situation of humans changes dramatically: they cease to be a measure of technical objects, whose actions become analogous and equivalent. The rule over the world, being the foundation of the idealistic design of the human, created in the Enlightenment Europe, conceived as apotheosis and glorification, is subject here to a deep erosion and becomes replaced by a so-called distributed agency. Böhle and Huchler address this matter by referring to German-language texts by Werner Rammert, but the notion of agency is one of the most fundamental concepts used by Bruno Latour, who encompassed all beings within it—including non-human beings, able to act, to function, "making some difference to a state of affairs, transforming some As into Bs through trials with Cs" (Latour 2005, 52).

Agency is no longer a pragmatic notion. It breaks the barrier of generality and abstraction, which elevates reasoning to a level of such entities as humanity or non-humanity, with the latter also including technical beings. Cyber-physical systems can be an object of a similarly advanced and abstract analysis. For instance, as the basis for the extrapolation of subsequent levels of the complexity of systems in the form of a system of systems (Bondavalli et al. 2016). But the notion, which gained an exceptional popularity in this area, describes a certain phenomenon occurring on the borderline between sociology, politics, and business—the so-called 4th Industrial Revolution. This is what Möller or Bartodziej, both quoted earlier, refer to. The former has no doubt that the last two decades, from the point of view of the year 2016, brought about a change in the digital revolution targeting industry, resulting in the emergence of cyber-physical systems, changing the relationship between humans and machines so dramatically.

To illustrate the essence of the change taking place, Möller makes use of a historical chart presenting the four subsequent revolutions: the first revolution, starting close to the end of the eighteenth century, the second revolution—dating back to the beginnings of the twentieth century, the third revolution—from the early 1970s, and the last—fourth—revolution, starting today, meaning together with the arrival of cyber-physical systems. The previous revolutions were also about grand ideas and technological inventions making it possible to turn these ideas into reality, aspiring to the notion of general purpose technologies (GPT). The first revolution took place with the use of the steam engine and railways, the second—appropriated mass production and electric power, the third—electronics and information and communication technologies (ICT). The chart Möller offers does not show the endpoints of revolutions. It rather suggests that one turns into another, continuing, lasting, overlapping each other as if in layers. With the passage of time, the rate at which they occur also grows; revolutions accelerate, appearing more and more often. They also become increasingly complex, which probably means not only the degree of their technical sophistication but also the extent of their impact on the surrounding environment, considering this factor in the context of cyber-physical systems.

Möller's chart is very similar to the idea published on the websites of the Davos World Economic Forum.[7] Unfortunately, the figure[8] explaining the sources of the adopted dating in detail, but in line with Möller's chart, is no longer available. These sources were based on specific inventions and indicated specific years: 1784 (establishment of the first spinning factory using the Watt steam engine), 1870 (Gramme machine), and 1969 (sending the first message over the? ARPANET, the prototype of the Internet). They were replaced by more detailed and advanced charts featuring more facts, mainly those concerning inventions—they can be found in a book by Klaus Schwab, the founder and executive director of the World Economic Forum (Schwab and Davies 2018, 10, 12). Most importantly, however, revolutions are presented there not in the form of points in time on a timeline, but as areas, the cross-sections of which change in the second dimension, forming thus entire complex areas. This second dimension illustrates a very challenging variable—the approximate human development index (HDI), but it is exactly this index, its nature, that brings us closest to the type of discourse we are dealing with. There are, however, more prerequisites to consider if we are to interpret this discourse correctly.

The pivot of the fourth industrial revolution are two books by Klaus Schwab: one written by himself only, from 2016 (Schwab 2016) and the other written in collaboration with Nicholas Davies, from 2018 (Schwab and Davies 2018). Both were published under the auspices of the World Economic Forum, and both are the canon reading for the idea which the Forum addresses at great length. It is one of the core ideas there, just like digital transformation. And while the latter is considered a certain mostly technological phenomenon, the fourth industrial revolution has gained the status of a superior and organizing construct, as illustrated, for instance, by the establishment of a special section of the World Economic Forum, named the Fourth Industrial Revolution.[9] It is a reversal of the hierarchy we have been using so far, and a very important piece of evidence of the type of story of the world as proposed by Schwab, clearly ideologized in its nature. It is a story of a competition-oriented, free-market economy which is the mechanism fueling the whole contemporary world, which means it is also political, a role of no lesser importance than the economic context. The Davos World Economic Forum is a perfect example of such a conjunction, and the recent manifesto, coinciding with the relatively near celebration of its 50th anniversary (it was founded in 1973), fits perfectly well in such a narrative1.[10] Its title is: *The Universal Purpose of a Company in the Fourth*

[7] https://www.weforum.org/ (accessed on: 1/2/2020).

[8] https://www.weforum.org/agenDA/2016/01/THE-Fourth-industrial-revolution-what-it-means-and-how-to-respond/ (accessed on: 2/23/2017).

[9] https://www.weforum.org/centre-for-the-fourth-industrial-revolution (accessed on: 1/2/2020).

[10] https://www.weforum.org/agenda/2019/12/davos-manifesto-2020-the-universal-purpose-of-a-company-in-the-fourth-industrial-revolution (accessed on: 1/2/2020).

Industrial Revolution.[11] The 'Company' mentioned in the title is something more than just a business organization; it is the salt of a capitalist earth. To make matters more interesting, the Forum was inaugurated by President Donald Trump, whose speech can be watched online.[12] This way, we reach a certain level of generalization of the subject of digital transformation, which, through other notions, revolving around themselves along with it, also makes the phenomenon a purely political event, where economic matters act as the foundation of the system.

It does not come as a surprise, therefore, that the European Commission addresses digital transformation individually, although it seems that the height of this enthusiastic interest dates back to 2018, the second year in a row when a document entitled *Digital Transformation Scoreboard* was published, discussing the progress of particular EU Member States in the implementation and adoption of digital technologies. It was a product of an EC-internal institutionalized project named Digital Transformation Monitor, which came to a halt, however, in mid-2018.[13] In 2019, a new kind of report was conceived, published under a more general name: *European Innovation Scoreboard.*[14] However, a separate chapter on digital transformation is found further, in the area of the Directorate-General, under the section of Internal Market, Industry, Entrepreneurship and SMEs department of the European Commission, and refers the reader to outdated websites from 2018. It seems the European Commission found the subject not as politically refreshing as only a while before. This does not, however, mean that digital transformation has become irrelevant. Rather, it means that it has lost its political charm and its ideological appeal in the eyes of European politicians. This, paradoxically, proves that it is treated instrumentally. However, the interest in digital transformation does not seem to be growing weaker among the providers and initiators of current economic discourses—audit/accounting firms. Each representative of the so-called Big Four, composed of: Ernst & Young, PricewaterhouseCoopers, KPMG, and Deloitte, have their own extensive package of information and guidelines. Ernst & Young speaks of eight main components of digital transformation, each of a different nature: cloud computing, robotics and intelligent automation, advanced analytics, digital strategy, next-generation data, Blockchain[15] technology, cyber risk, and digital architecture.[16] PricewaterhouseCoopers offers an own, metaphoric and descriptive strategy of a journey being digital transformation, where the most important question is one

[11] Ibidem.

[12] https://www.weforum.org/events/world-economic-forum-annual-meeting-2020/sessions/special-address-by-donald-j-trump-president-of-the-united-states-of-america-5e6010c5e7 (accessed on: 1/21/2020).

[13] https://ec.europa.eu/growth/tools-databases/dem/monitor (accessed on: 1/2/2020).

[14] https://ec.europa.eu/growth/industry/innovation/facts-figures/scoreboards_en (accessed on: 1/2/2020).

[15] A program solution enabling a dispersed control over variables and decentralized resources, eliminating the need for a centralized control center, used e.g. to maintain the consistency of the accounting records for virtual currencies such as Bitcoin.

[16] https://www.ey.com/en_gl/insurance/digital (accessed on: 1/2/2020).

about the fate of people divided into three categories: clients, communities, and leaders.[17] KPMG warns against considering digital transformation in overly simple terms and after an attempt to explain it, it offers a range of numerous and intricate services and technologies, assuming the role of a trustworthy and reliable guide through the maze so woven.[18] Browsing through Deloitte's website results in hundreds of reports on digital transformation, thus offering an impressive image of the company's expertise and versatility.[19]

Deloitte is also the co-author of a study that aims to appeal to the momentary and hasty fascination of its business clients, to a smaller extent at least, but also carries real analytical value. The study has been continuing for almost ten years now, and involves the participation of the MIT Sloan Management. It is called the Digital Business Global Executive Study and Research Project, and its recurring result is an annual insight into the situation of business organizations, focusing on their presence in the digital reality as seen by their employees, mainly those representing the management level. The study has been conducted since 2011 with a short break. Each year's edition sees a set of conclusions formulated in a different way, comprised synthetically by the changing team of researchers[20] in the title of the final report. Here is a handful of selected examples 2012: "Social Business: What Are Companies Really Doing? Social business is still just getting started. But its value is clearly emerging for marketing, innovation, operations and leadership"; 2013: "Social Business: Shifting Out of First Gear", 2015: "Strategy, not Technology, Drives Digital Transformation. Becoming a digitally mature enterprise"; 2016: "Aligning the organization for its digital future, digitally savvy executives are already aligning their people, processes, and culture to achieve their organizations' long-term digital success", 2017: "Achieving Digital Maturity: Adapting Your Company to a Changing World", 2018: "Coming of Age Digitally. Learning, Leadership, and Legacy"; or the most recent, 2019, tellingly succinct: "Accelerating Digital Innovation Inside and Out". The descriptive, or even anecdotal nature of these titles, combined with many years of studies, creates a clear narrative told as a historical tale. Digital transformation thus moves up one level of generality.

---

[17] https://www.pwc.com/us/en/about-us/workforce-strategy/our-digital-up-skilling-journey.html (accessed on: 1/2/2020).

[18] https://home.kpmg/xx/en/blogs/home/posts/2019/07/simplifying-your-digital-transformation.html (accessed on: 1/2/2020).

[19] https://www2.deloitte.com/global/en/misc/search.html#qr=%20Digital%20transformation (accessed on: 1/2/2020).

[20] Their latest team, working in the same composition for several years, is: Gerald C. Kane, Doug Palmer, Anh Nguyen Phillips, David Kiron, and Natasha Buckley.

Identity, Text, Knowledge

The abovementioned showcase of changes is, however, ultimately banal, and the organization of transformation will rather not mimic the deterministic model even though it may seem so at first glance—this is what the contemporary analyses of digital transformation deal with. Underneath, there are processes of a geological scale if we were to take a closer look at the notions used to describe them, getting thicker and thicker, welling up. They connect and form essential chains, with one of the first ones, quickest to work to the benefit of interpretation, and one of the most difficult ones at the same time, is culture. It is also a continuous changeability. It is a natural process of transformation of everlasting character, according to the Heraclitean principle of constant change. In the twentieth century, this changeability appeared particularly acute to Bauman, who called it *liquidity* and associated it with a historical phase he named *modernity* (Bauman 2000). Its randomness and variety has plunged humans into changeability felt as an inconvenience, difficulty, or even threat. Bauman discovers thus a principle of the reality, which appears to be hard to accept and come to terms with from a personal perspective, even though it has been known for a long time.

Experience exposes us to the effect of time almost incessantly. Experiencing and coming to terms with time is an age-old and trivial existential problem. Existence itself involves an inevitability of its dynamics based on time. Bauman considers the notion of liquidity in different variants, but all of them concentrate on the overwhelming experience of the melting, as he writes, of permanent structures and patterns concerning the various aspects of human life, in both individual and collective aspects. His book comes to being in the zenith of fascination with the vision of the world as a tangle of infinite relationships, where its fragment—one arranged from the available elements—can be seen only locally and for a brief moment, like a pattern seen inside a kaleidoscope. Inside of it, we can see a symmetrical and intricate order of colorful forms, but when we look into the light, a grain of bright pieces and tiles, it appears clear that this order is a mere illusion created by several reflecting surfaces. This experience teaches us about the impermanent, illusory, and local nature of our observations, even those most disciplined.

The same experience, but taking place at an anthropological and philosophical level, is a result of a violent storm that broke out in the first half of the twentieth century, ravaging our certainty of our knowledge of and about the world. It originates from the nineteenth-century crisis of mathematics, which by then had been an undisputed foundation of natural sciences. The said storm broke out more or less in the 1960s in the field of the humanities and social sciences, named differently in different disciplines: constructivism, postmodernity, poststructuralism, or postmodernism. It means confusion and uncertainty, and a need for new strategies. Not only scientific but also social ones. This is something strongly emphasized by Lyotard or Rorty. For Bauman, the consequences of the said storm reach eventually the everyday relationships established by humans as customers of institutions, practitioners of standards and duties, or participants of culture. In the face of this wind of change, all of the above, institutions, standards, duties, and culture, appear to be inconstant, random, usurper, volatile, and ambiguous. Changeable.

This melting pot also becomes the source of the phenomenon of identity. Bauman's book is published when the first wave of enthusiasm and fascination with the Internet reaches its peak, which is followed by March 2020—a time of a great stock market crash, which sweeps many collapsing companies investing in new technologies out of the market. The commotion caused by this fact is quite significant, and the crash itself and the speculation craze preceding it becomes known as *the dot-com bubble.*[21] Bauman, whose book appeared in 2000, could surely witness this American madness of investments in online innovations and the accompanying technologies, but the Internet is never mentioned in his publication. He addresses changeability and the problem of identity as civilizational problems of a quite different nature and a much more serious status. He carries on with an old subject raised ten years earlier by Giddens (1991), presented in his book in the form of a disturbing tale of the situation of the modern human near the end of the twentieth century. A human who loses their own sense of consistency and stability. Bauman's liquidity continues and acts next to technological changes, which seem to be dramatically important nowadays. In the twentieth century, identity is going through a crisis that is much more primary and fundamental.

This does not mean that the reflection on the human presence in a new environment, referred to then as a cybernetic environment, is absent whatsoever. Its most well-known exponent is Sherry Turkle, who has been studying the psychological foundations of this presence since the 1980s in her books, dealing with the relationships between humans and computers—as shown in *The Second Self. Computers and the Human Spirit* (Turkle 1984). In the mid 1990s, she published her probably best known work: *Life on the Screen. Identity in the Age of the Internet* (Turkle 1995). In line with the spirit of the times, she underlines the postmodernist ability of the Internet to create variable and multiple personalities. It lets people think of themselves as suddenly emerging characters, dispersed, multiple, flexible, in a continuous process of creation and, finally, fluid. Turkle uses the same expression as Bauman, but unlike him, she associates it with the Internet, which fits perfectly well in the same processes which Bauman views as broader, or—from a different perspective—deeper and more primary. She continues by arguing that the qualities she comments on are referred to by various and numerous types of reflections among which they travel freely, such as computer science, psychology, children's games, cultural studies, artificial intelligence, literary criticism, advertising, molecular biology, self-help, or artificial life. This means that her image of personality is commonly seen in different manifestations and inhabits the collective imagination of people exposed to the effects of the Web. Somewhere in the background, there is the looming image of a cyborg, with an own reflection—with its exponent being Katherine Hayles (1991).

---

[21] I have described this moment in detail in my book entitled *Pragmatyka internetu: Web 2.0 jako środowisko* (Maciąg 2013).

Katherine Hayles is also one of the many researchers engaged in the interpretation of the concept of the posthuman, which Turkle could not have yet considered in the mid 1990s. It was only 1999 when Hayles published her famous book entitled *How We Became Posthuman: Virtual Bodies in Cybernetics, Literature, and Informatics* (Hayles 1999). It is another variant of a similar narrative, based not on a psychological study, the generalization of which makes it possible to see and describe the reconstruction of personality and, further, identity, depending on the adopted perspective (which reminds of Mead's classical dualism of "me" and "I", the looking-glass self, based on the opinions of others, and the spontaneous self, the inner, individual response to these opinions respectively).

It is a narrative based on the observation of culture, the various texts bring the human into existence in the broad sense as some image, dream, goal, sense, etc. This stimulates an entire thought on the posthuman,[22] appearing as a new variant of the phenomenon in question, an answer to the world they live in, completely different from the world where they were brought to life as an idea by such philosophers as Descartes or Kant. Hayles belongs to those who see a change in the appearance of advanced computer improvements. She firmly believes that the so-called artificial intelligence is possible and almost at our fingertips, which urges us to solve, for instance, the problems of technical manipulation using our consciousness and thoughts. Today, this kind of belief has rather moved to the background, giving way to the growing gradual disappointment with the progress in the field in question, surfacing in the first decade of the twenty-first century, involving attempts to algorithmize knowledge by means of, for example, semantic networks. The numerous existing solutions of a similar kind do not claim the right to the range and impact considered by Hayles even though they do work. The enthusiasm of the late twentieth century in this area has experienced a grave disappointment.

Today's image of the human seems to be focused mainly on their network portrait, increasingly perfectly constructed from the data provided by humans themselves in various ways. Playing the game of trying on different identities, fascinated Hayles and many others alike—suffice it to recall Peter Steiner's famous image from 1993, showing a dog sitting in front of a computer and speaking to another dog sitting nearby, saying: "On the Internet, nobody knows you're a dog". The image was reprinted many times, and Steiner made a lot of money on it. But it has ceased to be valid too. Identity manipulation has taken a much more serious and much more covert form. Identity has become tangled in a network, emerging as a product—a product of a network, but one existing virtually, meaning thanks to social network sites[23] such as Facebook or Instagram.

In the foreword to Jennifer Golbeck's book entitled *Analyzing the Social Web* (Golbeck 2013), Ben Schneiderman writes that this is the first time in history in which humans get to use the tools to make social behavior patterns visible. What he

[22] I have covered this subject at length in in my book entitled *W stronę cywilizacji Internetu* (Maciąg 2016).

[23] We owe this expression to two authors: Danah Boyd and Nicole Ellison (2008).

means is the complicated relationships that connect humans with each other in a myriad of ways, forming a social network that evolves and changes continuously over time. It has been an object of sociometric studies since at least the 1930s. But studying these relationships is an unrewarding task. Even though they exist and clearly determine the fate of these people, they are invisible at the same time. Recreating them is tedious, imperfect, and prone to error. Yet, the problem disappears with the arrival of massive social networks materialized in electronic form. This way, they can be monitored, recorded, saved, and archived just as they occur in their great numbers and according to the time of occurrence. We can simply see how they come into being and change. In such a recording, every network portrait of every network node is clear, faithful, and reliable. As Aggarwal argues, the analysis of this data consists in detecting groups, in predicting the appearance of new relationships, estimating the range of the spread of information and the social reach of its impact, and in identifying experts (Aggarwal 2018). These are very specific subject matters that have emerged as highly interesting from the market perspective. Efforts to search and acquire this type of data became popular around 2010. Many books on the subject were published at that time. It is hard to wonder; researchers have entered their Garden of Eden: data about people is as available as never before, spreading before their eyes like boundless, growing, and ripe cornfields, ready to be harvested.

But this is not the only, recently discovered vault of information about people, considered both collectively and individually—and no longer able to hide in the crowd. The other treasure vault opens to the magic phrase of *Big Data,* and contains the first one, described a while ago. This is a certain concept and a marketing product at the same time. They shall not be confused; they are two faces of the same denotation: a huge set of any data. *Big Data* does not have a good definition in the ontological sense. Kitchin and McArdle write that the phenomenon emerges, acquiring traits resulting from its growing scale, which, as one can expect, describe a new and completely separate phenomenon (Kitchin and McArdle 2016). This intuition is really interesting. It could be compared to watching a glass of water and an ocean. The latter cannot be curbed, but opens an unlimited field of exploration, and its vast and rich expanse is a natural resource that is fought for. The case is similar with data, with the difference that the nature of the data can be of different type, meaning: of any type, depending only on the resourcefulness of the searcher. The sources of data can be really trivial and obvious, including credit cards, CCTV cameras, online traffic, cars with Internet modules, or remote energy meters, but there are also less obvious sources, which pertain to every digital trace left by a person, trackable and traceable by means of a growing number of digital devices surrounding us.

Kinoshita and Mizuno answer the question about what *Big Data* is as follows: "Big Data represent projections of things on real world, thinking of people, results of calculations of computer. To say concretely, they are numerical values, texts, images, movies, sounds, programs, and so on. They are coded by any rules, and stored in storages of computers. Computers can proceed any procedures to the coded things. The procedures are operations on binary digits essentially" (Kinoshita and Mizuno 2017, 92).

The world that emerges as a result of such a representation is strictly correlated with its prototype, bearing in mind that the translation of the real world into the digital world is not transparent and takes place according to a principle covered extensively in textbooks and discussed herein. As in the case of any translation, it is a reconstruction that is only similar to the source, performed in a different system of articulation of meanings, inaccurate. The colloquial meaning of translation is a materialization of this model as well because two different languages are, in fact, different systems of creation of meanings, and this is something quite well understood. In the first part, a lot of attention has been given to fundamental reflections on the matter, mainly those of Wittgenstein. The translation from the real world to the digital world, in the virtual sense, is based on the same mechanism but happens in different systems. The real world is a place where raw data is born, data based on physical and chemical processes related to the existence of organized matter. This organization existing by default and as a result of quite a basic conviction (since the attempts to understand it thoroughly—in quantum terms, for instance—remain unsuccessful) is a certain immanent trait of the real world. Humans would explain the sense of this organization in many ways, referring, for example, to the transcendental being of God, which had become the foundation of many religions, including Christianity. Defending this view has been fraught with bloody and cruel consequences. A completely different—secular—view, which has become the source of science until the present time, was offered by the already mentioned Galileo.

The real world is subject to translation into the digital world. Speaking more specifically, into a world subject to the pragmatics of a computer such as one designed by John von Neumann, as described in the introduction herein. In his IT textbook, a classical elementary guide to the logic and pragmatics of computer operation at the software level, Niklaus Wirth, already quoted in the book, directly emphasizes that the richness of the world needs to be reduced, simplified, that most of its aspects need to be ignored, and the elements that are really essential—adjusted to the computing capacity.[24] This is accompanied by a specific transition towards mathematical thinking because, essentially, the only symbolic system accepted by modern computers is the numerical system and mathematical modeling supported thereby. One could say that the modern computers like those we see to have developed turn Galileo's as well as Descartes' and Leibniz's visions into reality as they all dreamt of fact-based, mathematical systems of science and reasoning. But to them, mathematics was the essence of the (real) world, so applying it was simply a way to discover this essence. Today, the view has been questioned many times, as discussed before. This forces us to at least accept that mathamatics can be of a constructivistic nature. In light of this line of thought, this means that mathematics acts as a separate, perhaps artificial, system of meanings, different from the real world. In the practical application of computers, a translation that makes it possible to

[24] I discuss these process in more detail in my book entitled *Deus ex machina* (Maciąg 2012).

switch from one to another system is something common and obvious. And for this simple reason, one can insist on treating these devices as participating in the world only indirectly: by means of translation of what is real into what is digital, meaning mathematical and limited by the structural potential of very specific pieces of equipment at the same time. Only then we can see the duality of the digital world, especially of the world of *Big Data*. And this is the world an inforgs, observed by Floridi, lives in. Floridi addresses the matter as follows: "What I have in mind is rather a quieter, less sensational, and yet more crucial and profound change in our conception of what it means to be human. We are regularly outsmarted and outperformed by our ICTs. They 'reckon' better than we do. And because of this, they are modifying or creating the environment in which we live. We have begun to understand ourselves as inforgs not through some biotechnological transformations in our bodies, but, more seriously and realistically, through the radical transformation of our environment and the agents operating within it" (Floridi 2014, 94).

It needs to be added that at present, this is not only a theoretical concept but a real materialization of a digital *alter ego* of a living human. *Big Data* is not a project or a prediction, but a reality and practice.

The development of *Big Data* is, of course, of vital importance to the issues of subjectivity and identity considered here, especially in the context of virtual social networks. It follows two directions: a structured direction and a non-structured direction. The former, quite obvious, involves a development of the sociometric analysis, which in this case consists in a search for repeating patterns that enable one to discern a certain hidden organization of the social network, which leads further to predicting trends, analyzing preferences or habits, and establishing collaboration. The latter opens up to the vast richness that becomes the content of social relationships: of utterances, texts, movies, sounds. The advanced algorithms, based on neural networks, swarm intelligence, evolutionary processes, or unconventional logical systems are able to explore enormous amounts of data, creating models of their development, analyzing and predicting moods, and performing a semantic resorption of texts to the inside of digital systems, which then process and interact with them in ways known to humans only: they are able to respond to questions related to those texts or create their continuations while maintaining coherence therewith and—on top of that—enriching them. The latter property takes them to a world so far inaccessible to machines, meaning the world of language and its creations: texts, being a material for meanings and content, and not just records of symbols incomprehensible to machines: signs that can mean letters only for humans, arranged into sentences telling some story about the world.

The pair of notions: subjectivity and identity, mentioned earlier, denotes two fundamental aspects of anthropological perspective. This means looking at a human as a participant of a collective on the one hand, and as an individual with an internal power of self-determination and self-containedness on the other. Both are, of course, closely connected and mutually determinant. But the fragile balance between the two spheres has been stripped of mystery, hiding the human in the network of social interactions so far. The mystery was actually one of the foundations of human freedom, enabling humans to build the myth of their complexity and uniqueness. One of the most important philosophical ideas describing the sensitive and complex matter of this line of thought is the European tradition's notion of the other. The other, meaning the one beside me and bringing their own cosmos with them, just like me, with this cosmos revealed in countless and extremely sensitive incarnations and faces. Schneiderman's observation, recalled a moment ago, stresses the general property of the process which shines a bright light on the said mystery. The social human gets caught in their clear entanglement, which can now be modeled and predicted. In its entire overtness—colossal, growing at the speed of light—of the amount of data. Dens have been buried, forests have been cut down, landscapes have been mapped. There is nowhere to hide. A surprisingly fresh prediction in this context is offered by Foucault in 1966, according to which the human disappears, vanishes like a sand drawing on a beach (Foucault 1966). The elements of this metaphor aptly express the nature of the contemporary process leading up to the situation in question: the bareness of the occurrence of the trace, which is fragile precisely because it is caught in its unconcealed entity, bringing, of course, new problems regarding the interpretation of this "entity". This process adds yet another trait to this entity: its material, technical triviality. It also covers, let us add, the entire abundance of its creations, especially linguistic ones, subject to a similar operation. This is the process Floridi wrote about when he discussed his idea of the inforgs. But to him, in 2014, it was an emerging possibility. Six years later, it can be acknowledged as a currently ongoing materialization.

The main area of Foucault's said thought was the issue of knowledge, which he covered at length in an entire book, showing it mainly in the context of language. And this issue seems to be successfully continued nowadays. The ongoing reconstruction of subjectivity and identity boldly enters the area of language as a means of conveying sense and a reservoir of knowledge, but this time technically and directly. Language becomes an object of an advanced algorithmic analysis close to the end of the 1960s, and this is under the influence of the developing artificial intelligence. The idea seems obvious. Language is a reservoir of all content, which takes the form of organized disquisitions describing the world in its various aspects. Language is thus a door to the world of knowledge understood as a practical resource. The Platonic idea of knowledge, summarized in the slogan appearing in *Theaetetus* as a justified true belief, is gradually fading away. This definition, consisting of three qualities, is based on a certain human disposition, being conviction on the one hand, and refers to truth on the other. A truth which is the world's property when the definition is formulated. Knowledge, which is an expression of human initiative, exists thanks to its rooting in the world. This rooting is absolute and objective. Knowledge thus draws its life force from outside humans, to whom it, in fact, does not belong. It exists as a projection of the world, displayed only inside the human as a result of their special abilities.

At the beginning of the twentieth century, knowledge moves slowly along the line connecting the world and humans, or speaking more precisely: while acting as a connection between the world and humans, it changes its own direction. It gradually takes the freedom to enjoy access to the world away from humans, who hang on to its anchors tied firmly to the world, treating it as the source of their own certainty of their opinions concerning their own and the world's existence. An increasingly bolder thought emerges, expressed in a determined and cogent manner by Karl Mannheim in 1929, referring to it as the idea of the sociology of knowledge (Mannheim 1929). Knowledge is a product of social circumstances, where its bet against the truth is subject to increasingly rapid forces emerging in these very circumstances, leading eventually to a severance thereof. The sociology of knowledge, which is just being born, deprives it of its 'outerhuman' anchorage. It tears it away from the truth. This is only one step away from interpreting knowledge as a current resource, which happens rapidly and takes various forms in the second half of the twentieth century. Knowledge is used to interpret economic and political processes, it becomes a direct object of management in enterprises—equal to other resources and enables one to interpret the pragmatics of human activity. The most significant process concerning knowledge, initiated at the same time, but with its significance not clear yet, is probably the aspiration to formalize knowledge and thus to include logic into the operation of computers.

Such an opportunity arises in 1957 thanks to the groundbreaking and very famous book by Noam Chomsky (1957)—he interprets language as a machine generating sentences according to a finite number of principles contained in the human brain. Chomsky's achievement is unprecedented: he proposes an, in fact, formal way of combining human cognitive abilities with the world of linguistic practice. In order for this to succeed, he needs to separate the syntactic part, meaning the domain describing the principles of creation of linguistic structures, from the semantic part, i.e. the domain that gives meaning to the said structures. The former can be extracted from the practice of the use of language, with great reserves thereof being available to researchers. Discovering a syntactic order, an order which makes it possible for sentences and then—entire utterances—to appear, seems to be the first step towards the understanding of the mechanism that manages the principal process, which is about these sentences acquiring sense, and these utterances—the ability to accumulate information making up for the body of knowledge. Chomsky's idea becomes an inspiration for mathematical linguistics, and then for knowledge models of a formal—or actually mathematical—nature. They come into being within the just created and developing concept of artificial intelligence. Knowledge, which has actually appeared as a problem, is given a convenient and promising interpretation in the field in question. It becomes fixed in this area of scientific explorations.

The conviction underlying this development was as follows: "intelligence could be formalized as symbolic reasoning with explicit representations of knowledge" (Van Harmelen et al. 2008, vii), which meant simply a belief that such a representation is possible, but it seemed certain after Chomsky's revolution in the understanding of language. After all, language is the only system—or one of the most important symbolic systems—able to transmit, accumulate, or articulate knowledge. It should be added here that the discussion on the issue has been taking place not only here. The grounds thereof were formulated by Wittgenstein. It has been joined not only by linguists, but also other philosophers. In the 1950s, John Longshaw Austin gave a series of his famous lectures, where he dealt with colloquial language and its ability to influence the world. The lectures were then collected and published as *How to*

Do Things with Words (Austin 1962). They provide an impetus for a rapid exchange of sentences on the interpretation of roles played by utterances and texts. Their intensity and passion lead to the formulation of an observation concerning the transformation of the very method of philosophizing, which Rorty refers to as the linguistic turn (Rorty 1992). He thus underlines the significance given to language in terms of shaping the entirety of beliefs about the world, which can here be understood simply as the knowledge of and about the world. It is only a part of a great exchange of sentences, and of the multiplication of new views, focused on modern hermeneutics, on the ways in which all texts and utterances taking part in the astounding abundance language opens the door to function and act in practice. The dispute has continued in the field of the humanities since the 1960s, assuming names such as poststructuralism or postmodernism, also encompassing the area of historiosophic reflection (known under the name of postmodernity in Poland), bearing fruit in the form of works by Lyotard, Foucault, Derrida, and many others. Its greatest message, discovery, and concern at the same time is the way in which knowledge about the world is born, emerging as a substance resulting from a reaction caused only by historical, material, and social substrates. In this process, there is no metaphysical mysterious or transcendental sanction. There is only a reactive and experiential reality. A similar view can also be found in the domain of social sciences, for which this area is opened by, for instance, a famous work by Berger and Luckmann (1966), followed by a slightly later idea of frameworks by Goffman (1974).

This is how a tempestuous environment of discoveries concerning the nature of our relationship with the reality, which escapes the human cognitive ability and gets lost in artificial constructs, is born. The resource for these constructs is mainly language and its creations. This environment acts as a background to early, classical ideas of embedding knowledge into formal structures of a symbolic nature. Most of these early ideas appeared to be too simple to grasp the complex nature of knowledge; they referred, in fact, to simple concept structuring the semantic occurrences of language, which eventually always boils down to words and their surroundings. Collins and Quilian's semantic networks (Collins and Ross Quillian 1969), Minsky's frames (Minsky 1974), or Schank and Abelson's scripts (Schank and Abelson 1975) were based on an initial idea of semantic connections between notions and their recorded forms in the shape of a graph (network) which could be further complicated and detailed. The concept of ontology, similar to that of semantic networks, has stood the test of time most successfully (Sowa 2000, 2010, 2014). It is also the only one to reach a level of such fundamental deliberations, meaning that it has tried to arrange its foundations on a philosophical—metaphysical—level of analysis, getting into a debate with the meaning of the notion of ontology that can be found there (Poli and Seibt 2010).

However, this attempt to present knowledge, called a symbolic attempt (Flasiński 2016), did not manage to grasp the highly complex, dynamic, and ambiguous nature of emergence of the semantic language creations on a level above syntax. To put it in simpler terms, it failed to take in the variable and enigmatic semantic nature of language, which is its constitutive quality. For the sake of accuracy, it needs to be saod that attempts which Flasiński referred to as a "symbolic-numeric knowledge representation formulated in an explicit way" have also been made (Flasiński 2016, 225). They tried to solve the problem of ambiguity, vagueness, or changeability of the world by either selecting the conditions of reasoning appropriately or treating the input data accordingly. In the former case, it was assumed that knowledge was imprecise or incomplete (for example by applying Bayesian inference or non-monotonic logics). In the latter case, the assumption on the imperfection of knowledge was transferred as if earlier; by definition, it was assumed that it was impossible to 'reproduce' the world as it is, one-to-one (by applying the ideas of fuzzy and rough sets). It needs to be stressed that these historical solutions, especially the latter, belonging to the symbolic-numerical trend, have not vanished but have become a basis for various promising local applications (Brachman and Levesque 2004; Van Harmelen et al. 2008).

They are relevant to the discussion offered herein as a certain intermediate step towards the concept of accumulating knowledge in a formal way—meaning, simply speaking, with the use of computing procedures. This path leads to the abandonment of the principle of symbolism, which is a highly significant step in the elevation of the level of abstraction of the entire process. A very risky step, though. The idea of symbolic-numerical representation already poses a serious problem because it forces us to include mathematical structures upon which it is based into the discussion on symbolism. And it is a very fundamental discussion, which was already featured here when the history of mathematics was covered, and it seems to have no conclusive end. The abandonment of symbolism takes such a form of formalization—i.e. the accumulation and processing of knowledge where there is no clear connection between the structures that store and process this knowledge and the world of their denotations, meaning the reality they concern.

This is an approach which Flasiński calls computational intelligence, where knowledge is defined not explicitly, but implicitly. Starting from its most essential elements, this is the principle governing the mechanism behind deep neural networks where knowledge is kept dispersed in a great number of singular values of so-called weights and biases, assigned to particular neurons. In practice, these neurons are nothing more than organized sets of these values—complete sets of numbers recorded in mathematical structures. The connection of this type of knowledge with the external world has no semantic interpretation, just as in the case of symbolic representation. This type of a simple relationship does not exist hence it is impossible to semantically interpret the structure of a neural network, which remains a black box in this sense.

Although artificial neural networks have a longer history than solutions based on the symbolic representation of semantic language structures, they have been seen to develop at an increased rate only in recent years if this development is to be measured in the number of practical applications and the degree of access to the level of social awareness. The reasons behind it are named by François Chollet, an IT practitioner and designer of software used to create them: it is the substantial increase in computer performance, the improved algorithms, the development of the significance of and access to data, the increased expenditure on technology, and the more and more democratic access to technology in this area (Chollet 2017). The potential and popularity of artificial neural networks have made them an ever interesting 'hot topic' across all media. In the context of the problem of knowledge and language, we should focus our attention on only one—but really extensive—aspect of their development, which concerns a discipline known as natural language processing (NLP). Today, it is a large-scale area, with a long history, so it is hard to provide a good exhaustive description thereof. The challenge has been accepted by Deng and Liu (2018), who offer a definition by indicating, among others, its potential applications and organizing its development over time. They write in direct terms about the goal which is *to understand* the human, natural language. This innocent anthropomorphic figure seems to demonstrate the scale of the problem quite reliably. The applications include speech recognition, spoken language understanding, dialogue systems, lexical analysis, parsing, machine translation, knowledge graph, information retrieval, question answering, sentiment analysis, social computing, natural language generation, and natural language summarization.

Deng and Liu trace the beginnings of NLP back to the 1950s. They are based on a conviction that knowledge about language is embedded in the human brain and simply appears there as a result of deep inheritance. A major inspiration is the said idea of Chomsky, which leads to the discovery and composition of symbolic structures—also mentioned a while ago. This is a beginning of great significance: it introduces a pragmatic approach to language as a container of knowledge—one that can be recreated, which lets one gain access to this very knowledge. It is a turning point in the approach to knowledge. Knowledge starts being considered technically achievable and is thus seen as physical and phenomenal. Deng and Liu view this approach as the first wave of the process of digital utilization of knowledge. The second wave reaches its peak in the 1990s and abandons the fundamentalist, genetic view to turn to the empirical observation of language dynamics where the human brain operates based on the dominant sensory input material shaping its processes of association, pattern recognition, or generalization. Therefore, researchers make a shift towards huge sets of linguistic input data, to which they apply statistical procedures as well as early ideas from the area of artificial intelligence, including machine learning, which Deng and Liu refer to as "shallow".

The third stage of development of natural language processing, seen in the twenty-first century, triggers a revolution. It is based on the application of deep neural networks. Deng and Liu summarize its essence as follows: "Unlike shallow neural networks and related machine learning models, deep neural networks are capable of learning representations from data using a cascade of multiple layers of nonlinear processing units for feature extraction. As higher level features are derived from lower level features, these levels form a hierarchy of concepts" (Deng and Liu 2018, 7).

The most essential property of this idea is going straight to the semantic level, almost ignoring the syntactic level which previous approaches struggled with. The success of this transition, which opened the door to the amazing contemporary applications, is based on a simple idea, which solves a so far insoluble problem: how to make a machine act based on the semantic meaning of a text when processing it. How to make it overcome the barrier which humans handle with ease: to go beyond the symbolic, semiotic, and then syntactic levels, treating them only as an introduction to the level of sense. And then, of course, of knowledge.

A computer is an advanced abacus, which had until recently treated text as a sequence of symbols. Speaking in simpler terms, a sequence of graphic signs—letters, with arbitrary and random numbers assigned to them, forming a sort of code to become suitable for processing. These numbers did in no way reflect the possible semantic relationships formed by sets of letters arranged into words, sentences, or bigger textual units. The idea to overcome this limitation appears in a mature form in a text considered today as a historical beginning of a technology enabling the initiation of the formalization of meanings contained in texts. It was written by a group of researchers from Google, headed by Tomas Mikolov et al. (2013), and proposed an idea called *word embedding*. It was based on a quite simple and old observation according to which the meaning of a word in a text depends on the word's direct environment: other words surrounding it.[25] This observation, known as the distributional hypothesis, appears in texts from the 1950s: by Harris: "All elements in a language can be grouped into classes whose relative occurrence can be stated exactly (. . .) [...] it is possible to state the occurrence of any element relative to any other element" (Harris 1954, 146), and by Firth: "You shall know a word by the company it keeps" (Firth 1957, 11). Sandro Skansi, the author of a textbook on deep learning, calls references to Harris a "historical mistake" (he does not mention Firth at all, though). He refers, however, to Ludwig Wittgenstein and his already quoted *Philosophical Investigations*, naming the book the primary and most important source of the idea in question (Skansi 2018, 167).

The idea, in its simplified, practical form, involves observing and analyzing words that form a vicinity for the words we subsequently study. Of course, the studied words form a neighborhood for each other, which turns into a kind of a complex network because words in a text tend to repeat in many different—but not random—neighborhoods. The more text there is, the greater the number of such cases and the more precise the determination of the conditions of such a neighborhood, meaning the discovered semantic relationships. But there is still the remaining problem of how to organize such a network, and how to record it, formalize it, transform it into numbers in a way to make it suitable for computing. A tool able to do it all was provided by deep—multi-layer—artificial neural networks, which "learn" the network in question. The outcome of their application is a range of multidimensional spaces—the number of dimensions is unlimited, usually amounting to hundreds and thousands—where every word has a specific place, a certain 'spot' resulting from a certain specific set of values for each dimension. In other words, it has an assigned list of numbers whose order is significant because it reflects the order of the dimensions and is the same for each word.

[25] Such an idea only bears appearances of syntacticality, which shall be understood as a system, not as a physical proximity only.

But these numbers, which only appear to resemble their equivalents assigned arbitrarily, have a special property affecting their weight and making them unique: they determine the final meaning of words. This meaning is relative. Numbers describe the meaning of a word, but only in relation to other words. Words with a similar meaning gather into clusters in created spaces. But this is not all. The mathematical operations performed on the numbers describing these words result in a certain sense. For instance, the sum of numbers assigned to words "crown" and "woman" produces a mathematical result corresponding to the values assigned to the word "queen". Further stunning effects come to light, for example, when comparing such spaces, created for different languages: they are structurally similar, which means that words with similar or the same meanings in different languages occupy similar places in the space (they assume similar values for particular dimensions). A set of so defined words forms a universal language model, saved, recorded in an artificial neural network. And this is the only place where semantic relationships are included as lists of weights of neurons in its implicit layer. The individual dimensions do not have a semantic interpretation; they are only temporary scaffoldings, deprived of an own meaning, supporting a system of words prepared out of words. This way, the idea of knowledge accumulated implicitly, embedded in the very model and concealed in it, is materialized. As one can easily notice, the set of texts which is used as the basis for such a model should be as big as possible, and although there are different ideas to address this issue, a common practice is to use texts found across the WWW.

Mikulov and his associates, of course, acted inspired by observations and ideas of many of their predecessors. Publications regarding the same subject matter date back to the 1990s, becoming more and more numerous at the beginning of the twenty-first century (Young et al. 2017), illustrating the increasing popularity of the idea and the growing number of researchers exploring the subject. The most commonly mentioned publications include those by Glenberg and Robertson (2000), by Bengio and associates (2003), and by Collobert and Weston (2008). But what can be more interesting from our point of view is the further development of the idea described by Mikulov, which has gained an impressive momentum in recent years, which has translated into the emergence of algorithms of exceptional capabilities. The procedures of manipulating the source model of a language are becoming increasingly complex and refined, which involves a more and more ingenious manipulation of source numbers by incorporating additional criteria—for instance, by considering the semantic weight of words in their stream.

One of them is the so-called *attention technique*, where the value of words is determined based on each word's relevance defined separately, which is possible thanks to the application of a specific mechanism of memorization during the processing thereof. These are not entirely new solutions, but they evolve much faster and in a more efficient way from their simple to their advanced forms. In the case of the said technique, early forms of contextual analysis have been appearing since at least 2014 (Baroni et al. 2014) to evolve and become a more mature version of attention technique (Bahdanau et al. 2015; Luong et al. 2015; Vaswani et al. 2017). Next, they bear fruit in the form of extraordinary solutions, where a relevant text is easily generated by a machine as, for instance, a "made up" continuation of a narrative provided as the input. The machine is also able to answer indirect questions about the text, or translate the text with great accuracy. All these operations create a very deep illusion suggesting that a machine can understand texts like humans do (Devlin et al. 2018; Peters et al. 2018; Radford et al. 2019).

Transformation

In 1637, Descartes saw the publication of his *Discourse on the Method of Rightly Conducting One's Reason and of Seeking Truth in the Sciences* (Descartes 1637). The book's significance to the development of organized, disciplined, and thus scientific thought is very well known. It is also a work where the statement that has become one of the best known and most common philosophical maxims, repeated by Descartes himself, was featured for the first time: "I think, therefore I am". It emerged, as we know, from the practice of systematic doubt, which defended the mind against trusting illusions, which are a perfectly normal and known human experience. At the same time, this kind of thinking found the support only in humans themselves and their rationality, becoming expressed in poignant first sentences in a narrative that viewed the human as a fragile and care-needing owner of the world. This construct of human subjectivity, built for one hundred and fifty years, as Foucault claims in 1966, was disassembled subsequently by Nietzsche, Heidegger, and Foucault himself,[26] paving the way for new incarnations of humanism.

The reason for which the text was quoted here, being one of the most important initiatives of the said construct and one of the sources of its origin, comes from a place where these significant facts occur. This place is the space of rational thought, viewed with empirical accuracy and mathematical temperament. Descartes treats this human ability in very practical terms, as the path to knowledge. In the spirit of this practicality, Descartes, when speaking about humans, compares them with machines, offering proof for the impossibility to build an apparatus imitating the human, designed so well that one becomes unable to distinguish it from the prototype. In other words, he provides evidence proving that it is impossible to create such a machine. He offers two arguments to assert his claim: one is about the unique power wielded by humans over language, clearly highlighting that he does not mean the purely articulatory ability to utter words, but the contextual sense. He thus accidentally separates the layer of a pure language matter in which words and sentences clothe themselves to take the form of events from their internal order, which today we would call a semantic order. The other argument is about the unique ability granted to humans by the mind, enabling them to face any circumstances. Humans base their actions not—like automatic machines—on a predefined set of variants of behavior and of tools adapted to these variants, but on the power of consciousness

[26] I cover this in my book entitled *W stronę cywilizacji Internetu* (Maciąg 2016).

that can freely create and stimulate this behavior. Translating his words into terms closer to our understanding, Descartes means the process of inference based on accumulated knowledge, which means that the ability to learn must surely lie at the heart of this process. But Descartes' certainty, proving his and the humankind's exceptional and unique abilities, right at the emergence of this thought, in its earliest premonitions, is based on premises that have become quite outdated today.

Wisdom and knowledge are slowly but firmly leaving the exclusive territory of human authority, which seems quite banal in itself. What is of greater concern is the dynamics of this process, which maliciously follows several trends at the same time—and the reasons behind them have been collected here as articulations, effects, or coincidences, labeled collectively under the name of digital transformation. Their sources are dispersed and remote, but they are somehow obstinately closing in around one stubborn attractor.

The first trend emerges as a growing fundamental insecurity in the comprehension of the world and its nature. It is a paradoxical trend because it coincides with a bloom in the field of scientific studies and the range of their practical applications. It is accompanied by an increasingly excruciating, opportunistic process of instrumental treatment of everything, with knowledge at the forefront. The triviality of this trend stems from the predominant nature of the said instrumentalization dictated by economics and its crude pragmatics, referring to greed and egoism. It can make one sick, but it does solve the paradox. Come to think of it, this paradox seems inescapable. The instrumental approach, based only on this banal purpose, rebounds and loops back to itself as an egoistic benefit. In the context of the development of the digital economy, the process was dealt with by Benkler, who view the development of purely spontaneous communities as the foundations of a new, partnership-based, and selfless system (Benkler 2011). Yet, it was effectively counterbalanced by ordinary exploitation (Chandler and Fuchs 2019; van Dijck 2013; Fuchs 2016). But the described social process of instrumentalization of knowledge is only a practical materialization of a crisis of the processes of accumulation of knowledge (also in its scientific—credible and reliable—form), formalized into methodologies and the discourses arising out of them. The development of the relevant tools is accompanied by an increasingly complex image of the world available thanks to these tools—and a growing distrust of these tools at the same time.

The second trend pertains to humans in all of their roles; their metaphysical supremacy as exceptional beings, as unique social creatures, as estimable others, etc. All these grand and noble visions are being thoroughly reviewed. The human appears to be more of a naked animal, subject to an in-depth and permanent surveillance, but also one being modeled, satisfying current political or economic needs. It is, of course, a human who is no longer a mystery. Indeed, a truly unidimensional human. After the world lost its magical nature, it was only natural for the human to become exposed too, and it happened in the twentieth century at an increasingly consistent rate. It also turned out, as if coincidentally, that the design called human was a tool of both enslavement and discrimination, just like the broader cultural and political construct based upon them. One of the additional themes of the reconstruction of the role played by humans in history is the story of the notion of progress and of the belief in this optimistic process one should, as it appeared, be ashamed of on account of the terrible consequences caused by the development of science per se, especially of the disciplines: physics and chemistry.

And there is the third trend. The most important trend from the point of view of this book, which tries to associate a certain technique and a certain world of various events and processes with each other, guessing the hidden relationships existing between them. This association is expressed in different ways, the mechanical list of which was provided earlier. The most significant way, however, is the one that affects all others by taking over the technical methods of accumulating, storing, processing, and articulating knowledge, becoming a deep foundation of all and any changes. In accordance with the assumption of this book, quite unoriginal, to be honest, and well-proven in the literature on the subject, a relationship between language and knowledge exists, with the latter appearing to reach its fullest materialization in this relationship. And this very area, accompanied by spheres of other symbolic systems—e.g. those created in the domain of mathematics, is the fullest and most final territory where the processes of digital transformation occur.

This happens in line with the nature of complex phenomena, on all levels of analysis and in all areas at once, fulfilling—and this is what this alternation is about—the order of both the vertical application of analytical abstraction in the perspective of things, and of the horizontal extension of the area of essentially experiential primary research (observation). Each of these trends considered separately fails to grasp the process of transformation broadly enough to explain the local cause-and-effect processes, which surely exist, in a deterministic form, but on a level all too trivial for the phenomenon to be understood. This kind of approach needs to, perforce, refer to the highest level of description, which is knowledge itself, deprived of any basis transcending above it, which makes it simply unable to refer to any arbitrary criterion of its sense or credibility, including truth in the first place. But knowledge can be treated as an axiomatic system *par excellence*, and its inferences can make it possible to build a model which can be then tested and improved locally.

The reasoning covered above enables one to create an appropriate definition of digital transformation. This definition, careful, descriptive, may take the following wording: digital transformation is a space of knowledge, which is created by statements about it on the one hand and which shapes these statements on the other. The general ideas such as transformation, digitality, but also the more specific ones, such as e.g. cyber-physical systems, network, Industry 4.0, the 4th Revolution, complexity, and many, many more are all local subjects of these statements. This transformation takes place in time and is a set of trajectories of its components, interconnected with one another in a complex way, forming local clusters with their own trajectories with a range up to the level of the entire space inclusively. The most important, distinguishing quality of digital transformation in the historical context, allowed in the light of its temporal nature, is the phenomenon of the control of the said knowledge processes being slowly, but consistently, taken over by specialized machines.

References

Abolhassan, Ferri, ed. 2017. *The drivers of digital transformation*, Management for professionals. Cham: Springer International Publishing.

Aggarwal, Charu C. 2018. *Neural networks and deep learning: A textbook*. Cham: Springer International Publishing. https://doi.org/10.1007/978-3-319-94463-0.

Alur, Rajeev. 2015. *Principles of cyber-physical systems*. Cambridge, MA: The MIT Press.

Austin, John Langshaw. 1962. *How to do things with words: The William James lectures delivered at Harvard University in 1955*. Oxford: Clarendon Press.

Bahdanau, Dzmitry, Kyunghyun Cho, and Yoshua Bengio. 2015. Neural machine translation by jointly learning to align and translate. In *3rd international conference on learning representations, ICLR 2015, San Diego, CA, USA, May 7–9, 2015, conference track proceedings*. http://arxiv.org/abs/1409.0473.

Baroni, Marco, Georgiana Dinu, and Germán Kruszewski. 2014. Don't count, predict! A systematic comparison of context-counting vs. context-predicting semantic vectors. In *Proceedings of the 52nd annual meeting of the association for computational linguistics (volume 1: Long papers)*, 238–247. Baltimore: Association for Computational Linguistics. https://doi.org/10.3115/v1/P14-1023.

Bartodziej, Christoph Jan. 2017. *The concept industry 4.0. An empirical analysis of technologies and applications in production logistics*. Wiesbaden: Springer Fachmedien Wiesbaden GmbH.

Bauman, Zygmunt. 2000. *Liquid modernity*. 1st ed. Cambridge/Malden: Polity.

Bengio, Yoshua, Réjean Ducharme, Pascal Vincent, and Christian Janvin. 2003. A neural probabilistic language model. *The Journal of Machine Learning Research* 3 (null): 1137–1155.

Benkler, Yochai. 2011. *The penguin and the leviathan: The triumph of cooperation over self-interest*. New York: Crown Business.

Berger, Peter L., and Thomas Luckmann. 1966. *The social construction of reality*. New York: Doubleday.

Bharadwaj, Anandhi, Omar A. El Sawy, Paul A. Pavlou, and N. Venkat Venkatraman. 2013. Digital business strategy: Toward a next generation of insights. *Management Information Systems Quarterly* 37 (2): 471–482.

Böhle, Fritz, and Norbert Huchler. 2016. Cyber-physical systems and human action. A re-definition of distributed agency between humans and technology, using the example of explicit and implicit knowledge. In *Cyber-physical systems: Foundations, principles and applications*, ed. Houbing Song, Danda B. Rawat, Sabina Jeschke, and Christian Brecher, 1st ed., 115–127. Cambridge, MA: Academic.

Bondavalli, Andrea, Sara Bouchenak, and Hermann Kopetz, eds. 2016. *Cyber-physical systems of systems*, Lecture notes in computer science. Vol. 10099. Cham: Springer International Publishing. https://doi.org/10.1007/978-3-319-47590-5.

Bounfour, Ahmed. 2016. *Digital futures, digital transformation: From lean production to acceluction*, Progress in IS. Cham: Springer International Publishing. https://doi.org/10.1007/978-3-319-23279-9.

boyd, Danah M., and Nicole B. Ellison. 2008. Social network sites: Definition, history and scholarship. *Journal of Computer-Mediated Communication* 13: 211.

Brachman, Ronald, and Hector Levesque. 2004. *Knowledge representation and reasoning*. 1st ed. Amsterdam/Boston: Morgan Kaufmann.

Chandler, David, and Christian Fuchs, eds. 2019. *Digital objects, digital subjects: Interdisciplinary perspectives on capitalism, labour and politics in the age of big Data*. London: University of Westminster Press.

Chollet, François. 2017. *Deep learning with python*. New York: Manning Publications Co.

Chomsky, Noam. 1957. *Syntactic structures*. The Hague: Mouton.

Collins, Allan M., and M. Ross Quillian. 1969. Retrieval time from semantic memory. *Journal of Verbal Learning and Verbal Behavior* 8 (2): 240–247. https://doi.org/10.1016/S0022-5371(69)80069-1.

Collobert, Ronan, and Jason Weston. 2008. A unified architecture for natural language processing: Deep neural networks with multitask learning. In *Proceedings of the 25th International conference on machine learning*, ICML '08, 160–167. New York: Association for Computing Machinery. https://doi.org/10.1145/1390156.1390177.

Deng, Li, and Yang Liu, eds. 2018. *Deep learning in natural language processing*. Singapore: Springer Singapore. https://doi.org/10.1007/978-981-10-5209-5.

Descartes, René. 1637. *Discours de la methode pour bien conduire sa raison, & chercher la verité dans les sciences. Plus La dioptrique. Les meteores. Et La geometrie. Qui sont des essais de cete methode*. A Leyde: De l'imprimerie de Ian Maire.

Devlin, Jacob, Ming-Wei Chang, Kenton Lee, and Kristina Toutanova. 2018. BERT: Pre-training of deep bidirectional transformers for language understanding. *CoRR*. http://arxiv.org/abs/1810.04805.

Firth, John R. 1957. A synopsis of linguistic theory 1930-55. *Studies in Linguistic Analysis (Special Volume of the Philological Society)* 1952–59: 1–32.

Flasiński, Mariusz. 2016. *Introduction to artificial intelligence*. Cham: Springer.

Floridi, Luciano. 2014. *The fourth revolution: How the Infosphere is reshaping human reality*. Oxford: Oxford University Press.

Foucault, Michel. 1966. *Les mots et les choses. Une archéologie des sciences humaines*. Paris: Gallimard.

Fuchs, Christian. 2016. Critical theory. In *The international encyclopedia of political communication*, ed. Gianpietro Mazzoleni, 1–13. Malden/Oxford: John Wiley & Sons. https://doi.org/10.1002/9781118541555.wbiepc001.

Fuller, R. Buckminster. 1957. A comprehensive anticipatory design science. *Journal - Royal Architectural Institute of Canada* 34: 357.

Geisberger, Eva, and M. Broy. 2015. *Living in a networked world: Integrated research agenda cyber-physical systems (AgendaCPS)*. Munchen: Herbert Utz Verlag.

Giddens, Anthony. 1991. *Modernity and self-identity: Self and society in the late modern age*. Stanford University Press.

Glenberg, Arthur M., and David A. Robertson. 2000. Symbol grounding and meaning: A comparison of high-dimensional and embodied theories of meaning. *Journal of Memory and Language* 43: 379–401.

Goffman, Erving. 1974. *Frame analysis: An essay on the organization of experience*, Frame analysis: An essay on the organization of experience. Cambridge, MA: Harvard University Press.

Golbeck, Jennifer. 2013. *Analyzing the social web*. Amsterdam: Elsevier.

Hanna, Nagy K. 2016. *Mastering digital transformation: Towards a smarter society, economy, city and nation*, Innovation, technology, and education for growth. 1st ed. Bingley: Emerald Group Publishing Limited. http://search.ebscohost.com/login.aspx?direct=true&db=nlebk&AN=1423671&site=eds-live.

Harris, Zellig S. 1954. Distributional structure. *Word* 10 (2–3): 146–162. https://doi.org/10.1080/00437956.1954.11659520.

Hayles, N. Katherine. 1991. *Chaos and order: Complex dynamics in literature and science*, New practices of inquiry. Chicago: University of Chicago Press.

———. 1999. *How we became Posthuman: Virtual bodies in cybernetics, literature, and informatics*. Chicago/London: University of Chicago Press.

Howe, Neil, and William Strauss. 1991. *Generations: The history of America's future, 1584 to 2069*. New York: William Morrow.

Kinoshita, Eizo, and Takafumi Mizuno. 2017. What is big data. In *Big data management*, ed. Fausto Pedro García Márquez and Benjamin Lev, 91–101. Cham: Springer International Publishing. https://doi.org/10.1007/978-3-319-45498-6_5.

Kitchin, Rob, and Gavin McArdle. 2016. What makes big data, big data? Exploring the ontological characteristics of 26 datasets. *Big Data & Society* 3 (1): 1–10. https://doi.org/10.1177/2053951716631130.

Latour, Bruno. 2005. *Reassembling the social: An introduction to actor-network-theory*. Oxford: Oxford University Press.

Leignel, Jean-Louis, Thierry Ungaro, and Adrien Staar. 2016. *Digital transformation information systems governance*. London/Hoboken: ISTE, Ltd./Wiley.

Leodolter, Werner. 2017. *Digital transformation shaping the subconscious minds of organizations*. Cham: Springer International Publishing. https://doi.org/10.1007/978-3-319-53618-7.

Luong, Minh-Thang, Hieu Pham, and Christopher D. Manning. 2015. Effective approaches to attention-based neural machine translation. *CoRR*. http://arxiv.org/abs/1508.04025.

Maciąg, Rafał. 2012. *Deus ex machina: nowe media i ich projekt poznawczy*. Kraków: Wydawnictwo Uniwersytetu Jagiellońskiego.

———. 2013. *Pragmatyka Internetu: Web 2.0 jako środowisko*. Kraków: Wydawnictwo Uniwersytetu Jagiellońskiego.

———. 2016. *W stronę cywilizacji Internetu: zarządzanie w naukach humanistycznych*. Kraków: Wydawnictwo Uniwersytetu Jagiellońskiego.

Maedche, Alexander, Jan vom Brocke, and Alan Hevner, eds. 2017. *Designing the digital transformation: 12th international conference, DESRIST 2017, Karlsruhe, Germany, May 30—June 1, 2017, proceedings*. Cham: Springer.

Mannheim, Karl. 1929. *Ideologie und Utopie*. Vol. 3. Bonn: F. Cohen.

Matt, Christian, Thomas Hess, and Alexander Benlian. 2015. Digital transformation strategies. *Business & Information Systems Engineering* 57 (5): 339–343. https://doi.org/10.1007/s12599-015-0401-5.

Maturana, Humberto R., and Francisco J. Varela. 1992. *The tree of knowledge: The biological roots of human understanding*. Revised ed. Boston/New York: Shambhala.

Mikolov, Tomas, Ilya Sutskever, Kai Chen, Greg Corrado, and Jeffrey Dean. 2013. Distributed representations of words and phrases and their compositionality. *ArXiv* (October) http://arxiv.org/abs/1310.4546.

Minsky, Marvin. 1974. *A framework for representing knowledge*. Cambridge, MA: Massachusetts Institute of Technology.

Möller, Dietmar. 2016. *Guide to computing fundamentals in cyber-physical systems: Concepts, design methods, and applications*. Cham: Springer International Publishing.

Peters, Matthew E., Mark Neumann, Mohit Iyyer, Matt Gardner, Christopher Clark, Kenton Lee, and Luke Zettlemoyer. 2018. Deep contextualized word representations. *CoRR*. http://arxiv.org/abs/1802.05365.

Poli, Roberto, and Johanna Seibt, eds. 2010. *Theory and applications of ontology: Philosophical perspectives*. Dordrecht: Springer Netherlands. https://doi.org/10.1007/978-90-481-8845-1.

Radford, Alec, Jeffrey Wu, Rewon Child, David Luan, Dario Amodei, and Ilya Sutskever. 2019. Language models are unsupervised multitask learners. February. https://d4mucfpksywv.cloudfront.net/better-language-models/language-models.pdf.

Rorty, Richard M., ed. 1992. *The linguistic turn: Essays in philosophical method*. Reprint ed. Chicago: University of Chicago Press.

Salkin, Cere, Mahir Oner, Alp Ustundag, and Emre Cevikcan. 2018. A conceptual framework for industry 4.0. In *Industry 4.0: Managing the digital transformation*, Springer series in advanced manufacturing. Cham: Springer International Publishing. https://doi.org/10.1007/978-3-319-57870-5.

Schallmo, Daniel R.A., and Christopher A. Williams. 2018. *Digital transformation now! Guiding the successful digitalization of your business model*, Springer briefs in business. Cham: Springer International Publishing.

Schank, Roger C., and Robert P. Abelson. 1975. Scripts, plans, and knowledge. In *Proceedings of the 4th international joint conference on artificial intelligence – volume 1*, IJCAI'75, 151–157. San Francisco: Morgan Kaufmann Publishers Inc. http://dl.acm.org/citation.cfm?id=1624626.1624649.

Schwab, Klaus. 2016. *The fourth industrial revolution*. New York: Crown Business.

Schwab, Klaus, and Nicholas Davies. 2018. *Shaping the fourth industrial revolution*. Geneva: World Economic Fund.

Skansi, Sandro. 2018. *Introduction to deep learning: From logical calculus to artificial intelligence*, Undergraduate topics in computer science. Cham: Springer International Publishing. https://doi.org/10.1007/978-3-319-73004-2.

Sowa, John F. 2000. *Knowledge representation: Logical, philosophical, and computational foundations*. Boston: P.W.S.

———. 2010. The role of logic and ontology in language and reasoning. In *Theory and applications of ontology: Philosophical perspectives*, ed. Roberto Poli and Johanna Seibt, 231–263. Dordrecht: Springer Netherlands. https://doi.org/10.1007/978-90-481-8845-1_11.

———. 2014. *Principles of semantic networks: Explorations in the representation of knowledge*. Morgan Kaufmann.

Stolterman, Erik, and Anna Croon Fors. 2004. Information technology and the good life. In *Information systems research: Relevant theory and informed practice*, ed. Bonnie Kaplan, Duane P. Truex, Wastell David, A. Trevor Wood-Harper, and Janice I. DeGross, 687–692. Boston: Springer US. https://doi.org/10.1007/1-4020-8095-6_45.

Torngren, Martin, Fredrik Asplund, Bensalem Saddek, John A. McDermid, Roberto Passerone, Holger Pfeifer, Alberto Sangiovanni-Vincentelli, and Berhard Schätz. 2016. Characterization, analysis, and recommendations for exploiting the opportunities of cyber-physical systems. In *Cyber-physical systems: Foundations, principles and applications*, ed. Houbing Song, Danda B. Rawat, Sabina Jeschke, and Christian Brecher, 1st ed., 3–14. Cambridge, MA: Academic.

Turkle, Sherry. 1984. *The second self: Computers and human Spirit*. London: Granada.

———. 1995. *Life on the screen: Identity in the age of the internet*. New York: Simon & Shuster Paperbacks.

Ustundag, Alp, and Emre Cevikcan, eds. 2018. *Industry 4.0: Managing the digital transformation*, Springer series in advanced manufacturing. Cham: Springer International Publishing.

van Dijck, José. 2013. *The culture of connectivity: A critical history of social media*. Oxford/New York: Oxford University Press.

Van Harmelen, Frank, Vladimir Lifschitz, and Bruce Porter, eds. 2008. *Handbook of knowledge representation*. Amsterdam/Boston: Elsevier.

Vaswani, Ashish, Noam Shazeer, Niki Parmar, Jakob Uszkoreit, Llion Jones, Aidan N. Gomez, Lukasz Kaiser, and Illia Polosukhin. 2017. Attention is all you need. *ArXiv* (June) http://arxiv.org/abs/1706.03762.

Williams, Bill. 2012. *The economics of cloud computing*. 1st ed. Indianapolis: Cisco Press.

Wirth, Niklaus. 1976. *Algorithms + Data structures=programs*. Englewood Cliffs: Prentice-Hall.

Young, Tom, Devamanyu Hazarika, Soujanya Poria, and Erik Cambria. 2017. Recent trends in deep learning based natural language processing. *ArXiv* (August) http://arxiv.org/abs/1708.02709.

Part II
Knowledge

Chapter 3
Wittgenstein and the Beginnings of Adventure

The technology usually associated with digital transformation has been developing for dozens of years now. There is a clear demand for a synthetic judgment of the reality in which the said technology has developed. Let us, however, avoid a phrasing which describes the relationship between technology and reality in overly simple terms, as a unidirectional influence. It is a lesson into the questionable value of the ordinary technological determinism—a lesson already learned. If we reject it, we open a door to new ways of understanding, ones we are going to look for here. And we will certainly become entangled in an analysis of naming as a certain significant historical process, which is not as much linguistic as cognitive in its nature—out of necessity, as if, since it is a tale of the effort humans put into describing the world using words.

A thus formulated necessity carries an obvious decision in it; it does not do with a silent approval of a separate, sovereign existence of two different realities: world and language, names, naming in general. Such an assumption is usually made at the beginning of a great number of excellent research papers, which imply that the world is examined as it is, that there is a direct insight into its content, and the procedure followed to perform this examination only has a technical meaning. It comes down to the selection of the right narration technique, a well-defined set of tools (such as notions), and a coherent cognitive system to glue everything together. Such systems can be found in a scientist's 'storehouse', in compartments labeled with the names of individual scientific disciplines. Naming is a part of the said set of tools.

Such a procedure is basically correct and the only right way to proceed taking the objective being to describe the world into consideration, respecting its diversity on the one hand, and trying to keep a cool distance on the other, expressed in the technical language suited to the object of examination. And yet, we can feel there are some concessions in this modus operandi, a certain failure, a certain mistake. This sensation becomes all the more painful when the selected coherent cognitive system drifts apart from the world. For instance, it has no chance to become reinforced in the domain of experience. How can one make philosophical analyses well rooted in

R. Maciag, *Knowledge as a Tale*, https://doi.org/10.1007/978-3-031-69820-0_3

experience in the first place? Or philological analyses? Some believe that in such circumstances, the system loses the right to be called a science. When the language used to speak about the world becomes hermetic, we, not belonging to the group of its users, feel cheated since we are deprived of access to the discovered world, which violates the ancient unspoken principle according to which the world belongs to everyone. Every research process must therefore return to the natural language to become a social property. Moreover, since a research process is usually a collective effort, it is governed from the very start by social processes, as observed and described by a Polish doctor of Jewish origin, Ludwik Fleck in 1935 (Fleck 1935). The immersion in social processes means an immersion in the reality of a language that needs to weaken in its exactitude, accuracy, and explicitness while becoming more common, general, and understandable. In the case of very common and global phenomena, these problems tend to be urgent.

Digital transformation is a certain assumption concerning the world's phenomena and a name at the same time. And if we were to follow the ordinary way of separating phenomena and words in keeping with the notional discipline, we could refer to it as if to some specific thing, defining this thing beforehand. And this is what most researchers do, in fact. Those, who, as it appears, see digital transformation as a phenomenon possible to be described using notions from the domain of management and economics. The matter will be addressed later on. There are those who view it as a tale of technological development. Out of necessity, such tales feature an important social context. This gives birth to quite a pleasant and extensive field of description. But the premise of this book is based on a premonition that there is something more to it. It is a very free premonition, which sees a symptom of very deep changes in the process taking place in the world of today. These changes can be called digital transformation. It can also be assumed that "digital transformation" is only a random group of words, which has come to being as a result of deeper-level linguistic processes which we commonly witness in the form of naming. And since these linguistic processes are in fact signals of cognitive processes, they are the expression of a certain knowledge about these said changes. And their range and scale make this knowledge highly significant.

The appearance of the latter word—knowledge—means trouble. This is because it opens the door to one more space which we usually set in opposition to the world. Which is, of course, a mistake because knowledge, whatever it may be, is a part of this world. But it is necessary if we wish to maintain the discipline of speaking about the world in terms of a separate fragment of reality, lying in front of us. Willard Van Orman Quine has gone to great lengths to discuss the problem in question. The starting point of his discourse is shamelessly blunt: "I am a physical object sitting in a physical world. Some of the forces of this physical world impinge on my surface. Light rays strike my retinas; molecules bombard my eardrums and fingertips. I strike back, emanating concentric air-waves. These waves take the form of a torrent of discourse about tables, people, molecules, light-rays, retinas, air-waves, prime numbers, infinite classes, joy and sorrow, good and evil." (Quine 1957, 1).

Hence, education of language creates knowledge which makes the survival of not only humans but also animals possible. Animals are, however, abandoned on the way to an increasingly advanced development of language, which eventually adopts subtle semantic and logical mechanisms (Quine 1995).

Quine is driven by the good-old Platonic idea of knowledge: in the end, it is the human who possesses it. Yet, a fraction of knowledge, a substitute thereof is also held by animals, but treated as separate beings, individuals, with each of them equipped with a perceptual apparatus and tools to process the perceptual stimuli like humans do. Let us, however, abandon this perspective for certain valid reasons. And what if we deprived humans of the sanction that makes it possible for knowledge to exist? This measure has a lot to do with digital transformation, which—given the technical innovations existing in the area of the so-called artificial intelligence—needs to review the position of knowledge in the world, and detaches itself from acknowledging human exclusivity in this domain. Machines collect, store, and process knowledge as well For example, machines that belong to the branch of technology referred to as artificial intelligence, but not only (Burgin 2015; Minsky 1974; Van Harmelen et al. 2008; Weller 2010). The problem will be addressed in more detail in the final part of the book.

In a game played with the world, the ability to describe the latter, aptly analyzed by Wittgenstein, has to take a form based on an experience of an encounter with richness and plenty—the first experience thereof is the great number of experienced phenomena and the mystery of their coexistence. This experience gives rise to the idea of combination, which will eventually appear to be a descriptive figure of fundamental significance. "because an account is essentially a weaving together of names", says Socrates as depicted by Plato (1987, p. 116), which is eagerly yet inaccurately—and seemingly purposefully—cited by Wittgenstein in his *Philosophical Investigations*: "For the essence of speech is the composition of names"(Wittgenstein 1958, 21). This inaccuracy involves changing the effect the composition of words (names) causes. What changes is the area which the original text describes very generally: *einai logou ousian* (ειναι λογου ουσιαν), which can be translated as gaining a meaning or a rational account (Chappell 2005, 202). Here, Wittgenstein refers to speech, changing and specifying Socrates' idea to a considerable degree. Although Wittgenstein does mention names, thus referring clearly to language, which he also stresses, he quickly moves on to make the world, the reality the object of his considerations. In the sentence coming right after the quote from Plato, he recalls "simple constituent parts", meaning primary elements, which are the basis of combination of words. In order to describe them, he refers to Russell's idea of "individuals" or to his own concept of "objects" (Wittgenstein 1974, 7). These small elements, the building blocks of the world, form combinations, structures, compounds, relationships. The world is obviously an outcome of combination.

Wittgenstein's placing of the name of Russell next to his own in this context is by no means a coincidence. Russell was his mentor and teacher. He described his idea of "individuals", clearly a source of inspiration for Wittgenstein, quite early—in a text from 1911, explaining at the same time the ontological status of these beings: "I believe there are simple beings in the universe, and that these beings have relations in virtue of which complex beings are composed" (Russell 2003, 94).

Since the relationship between the world and thoughts is governed by logic, it is possible to avoid the problem resulting from collating reality with language. Logical beings are of a different kind, as if more primary, more fundamental. The order of logical beings shapes the order of the world as accessible to thoughts. Language is only the outcome of this operation.

The smallest atom of logic is a term. Terms can take two forms: terms themselves and their relationships or predicates describing them. Russell calls the former *particulars*, and the latter—*universals* (Russell 2003, 95). Examples of the latter include "diversity, causality, father, white". *Particulars* are the part of a being, which is not a relationship or a predicate, meaning not what gives the being its quality and position. The concept of *particulars* will be utilized later by Armstrong in his theory of ontology (Armstrong 1997), which will grow to become a foundation of a broader idea of knowledge (Maciag 2022).

This original concept, elevating logic to the highest rank of a metaphysical, universal, ideal platform of the world's existence, had evolved over time, taking on an increasingly advanced form (Klement 2016). It led to a final and very interesting conclusion that the knowledge of those components—*simples*—was not necessary to acquire knowledge about the world, also in the meaning of the ideal Omniscience: "We could distinguish in the world a stuff (to use William James's word) and a structure. The stuff would consist of all the simples denoted by names, while the structure would depend on relations and qualities for which our minimum vocabulary would have words. This conception can be applied without assuming that there is anything absolutely simple" (Russell 1948, 276).

This reasoning is far from easy, contrary to appearances. The world is most likely generally complex and non-simple. The presence of the simplest components is questionable and unnecessary to understand it. They need to be discarded, which seems paradoxical because this violates the logic of totality and its parts. The relationships that can exist autonomously are therefore the most important and decisive, in fact.

In his *Tractatus Logico-Philosophicus*, Wittgenstein offers an extensive description of the way the world exists, using slightly different notions of fact and the state of affairs generating such facts. The description comes from a broader task, which is, as in the case of Russell, to understand the relationship combining a thought with the world, meaning a certain human ability of cognitive nature but viewing the role of language differently. It addresses a fundamental and ancient philosophical task: to answer the question of the relationship between human knowledge and the world; in other words, the question of what we actually know. This gives rise to a further, more specific question about the truth or the value of science.

Wittgenstein published his *Tractatus* in 1921, but presented his first draft of *Notes on Logic* to Russell already in 1913 (he became his student in 1912, aged 23) (Wolniewicz 2000). Their exchange of ideas as seen in the texts is clearly a result of their actual collaboration, although Wittgenstein, seventeen years younger than Russell, is formally a student in this relationship.

Wittgenstein analyzes the mutual relationship of the world and language by following the basic steps of reasoning in the form of ordered and numbered propositions. From the point of view of this text, the most important selected propositions, forming a consistent sequence of inference, are: "The world is the totality of facts, not of things. (…) A logical picture of facts is a thought. (…) A thought is a proposition with a sense. (…) The totality of propositions is language" (Wittgenstein 1974, 5, 12, 22). They describe the dependence of subsequent levels or areas bonding the world, thoughts, and language into a whole. They include, therefore, a solution to the problem emerging here as a mutiny against the persistent juxtaposition of two extremes: the world and language. There is a clear transmission between them. One composed of well-defined intermediaries: facts, logic (*logical picture*), thoughts, and sentences (*propositions*).

The least obvious 'mediator' among them is logical order—the one that determines all and any conditions of thoughts taking form. The idea of a logical picture refers to the structure of logical space. It is a subject common to both Russell and Wittgenstein. It is definitely not a friendly space; it requires an additional assumption and a description. Wittgenstein argues: "the facts in logical space are the world" (Wittgenstein 1974, 5). The space of facts should never be confused with the space of logic. Their connection is Wittgenstein's grand idea to answer the question about the relationship between thoughts and the world. Wolniewicz depicts this relationship using the following words: "the world is an island of facts (…) in an ocean of possibilities, being this very space" (Wolniewicz 2000, XV). A thought accesses the world through the said space, and accesses this space through sentences, which are a creation of language, in turn. In his *Tractatus*, Wittgenstein goes to great lengths to show the mechanism of the existence of sentences enabling the fulfillment of the said role. A sentence becomes the most important structure making it possible, on the one hand, to utilize the potential of logic and to represent the world on the other. Yet, the logical space interwoven with the world of facts remains a mysterious, isolated place. A place where the obvious affiliation to either of the parties—the world and man—becomes blurred. There is no telling where logic comes from and who makes it what it is.

In order to describe the mutual relationship between the said parties (the world and humans), Wittgenstein needs additional levels of explanation and additional instances such as fact, meaning a system of objects, logic, a thought, a sentence, a language. In the introduction to the first edition of *Tractatus*, Russell referred to it directly as the first step towards the theory of knowledge (Russell 2002, ix). Thinking and its product, a thought,—a human ability—is surely an instrument of power, like a sentence and a language. The tradition was originated by Plato in a well-known definition featured in *Theaetetus*: *Dóksa alethés metá lógu*, translated by Waterfield into English as: "true belief accompanied by a rational account" (Plato 1987, 115), usually shortened to *justified true belief* (Dancy et al. 2010). The definition requires the presence of the one giving the account. This way, a situation of fundamental significance to knowledge, one defining it, is created. A situation which certainly involves the participation of humans. For Wittgenstein, a language, a sentence, a thought are like a human-bound fragment of transmission taking place between the human and the world. Facts are certainly world-bound, in turn. Between facts and thoughts, there is a mysterious, borderline space of logic.

In the said definition and in the citation Wittgenstein referred to, the same vital notion of 'logos' (λόγος) appears to highlight the significance of combining names with one another, as mentioned earlier. The notion has a long tradition of interpretation, but it seems to be understood in a quite pragmatic and simple way in the fragments of interest to us. The fact is illustrated by the various translations of the notion: *das Wesen der Erklärung* (Plato 1856, 208), *rational account* (Plato 1987, 115), *l'essence de leur definitione* (Deschoux 1980, 289), *ścisłe ujęcie* (Platon 2002, 178). Each deals with a certain technical property of a situation, related to the order and discipline of reasoning. What is surprising, though, is the use of the same notion in the case of the definition of knowledge, and the effect produced by the interweaving of names to which we referred earlier. This coincidence unexpectedly reinforces and intensifies the latter case. As if the interweaving of names had a lot in common with knowledge, made it possible to some extent, or at least kept some 'kinship' with it. The interweaving of names is not therefore a purely rhetorical measure; through some bizarre ability to create relationships, it becomes elevated to the cognitive level. Names are neither gentle nor passive; when they get too close to each other, like some radioactive isotopes, it may lead to an explosion—in this case, an explosion of semantic consequences resulting in the image of the world being created in language. This mutual approximation denotes relationships, mutual impact, interaction, and other effects exerted on each other by sets of reagents, which cannot remain stable, dormant.

But Wittgenstein does not cease at the level of understanding of the way language exists in the context of the world he proposes in his *Tractatus*. One of the main domains where it can be found defied is logical order. In *Philosophical Investigations*, Wittgenstein dismantles the structure in question as, in essence, a weakness of reasoning. Introducing the idea of a logical picture and acknowledging logic as a boundary and critical area at the same time was continued in Russell's opus magnum entitled *Principia Mathematica*, co-written by Alfred Whitehead and published in the form of three volumes in the following years: 1910, 1911, 1912. This was based on an assumption that a universal logical system laying the foundation to mathematics existed and through it—perhaps even to the entire science. The twentieth century is more of a landscape of proliferating views according to which the search for such a fundamental, transcendental order, able to capture the world in a succinct abstract of a great formal system of a transparent and coherent structure is a pipe dream. The most famous—and deadly—blow was dealt to this conviction by Gödel in 1931, in his well-known text, the title of which is a direct reference to Russell and Whitehead's work (Gödel 1931). When commenting on Gödel's text, Nagel and Newman argue: "no final systematization of many important areas of mathematics is attainable, and no absolutely impeccable guarantee can be given that many significant branches of mathematical thought are entirely free from internal contradiction" (Nagel and Newman 1958, 6). Wittgenstein's *Philosophical Investigations* reveal a rapid turn, which involves rejecting the logical, stable, transcendental structure governing thoughts, sentences, and language in favor of a free interplay. The idea opens the text and the analysis thereof continues for many pages, making the author's imagination clearly run wild. It offers a vision of a language that is lively and complex but also disorganized and random: "Our language can be seen as an ancient city: a maze of little streets and squares, of old and new houses, and of houses with additions from various periods; and this surrounded by a multitude of new boroughs with straight regular streets and uniform houses" (Wittgenstein 1958, 8).

In such a language, names cannot be subject to transparent structural principles; they become an object of temporary discovery, determined by the circumstances of their emergence, by the interplay of contexts: "the speaking of language is part of an activity, or of a form of life" (Wittgenstein 1958, 11). The word: life takes the place of the crystal-clear logical structure, which in *Tractatus* consumed and enveloped the world, being only a small, local, chosen manifestation of its potential. In the metaphors featured in *Investigations*, the situation is reversed: now, the world, through its mysterious order and dynamics, establishes local names which cannot be stable since they stem from the intertwining of circumstances which are different each single time again and again. Transcendentalism yields to naturalism, which opens the pathway to the interpretation of the nature and the dynamics of the said circumstances. This will be dealt with in the second half of the twentieth century by the language theory—and with great enthusiasm as shown in the works by such researchers as John Longshaw Austin, Michel Foucault, Jacques Derrida, Richard Rorty, and many more.

Moving from logic to life as spaces determining the way in which names and sentences describing the world can be considered a revolution, but it may as well be interpreted as a supplementation and an amendment made to the ontological structure proposed in *Tractatus*. In the space between the world and the human, there are still mediating linguistic structures such as a sentence, and only the conditions of formulating them and their meanings change, or –more precisely—multiply. Wittgenstein resigns from the only universal logical structure to opt for multiplicity and variability. And yet, first, it seems that he does not oppose the former by turning towards the latter; the logical structure is relocated from the only dominant position to the area of many potential possibilities. Secondly, the said multiplicity and variability, being the effects of the complex dynamics called life, do not rule out order. The only difference is that this order needs to be of a different kind.

At the time of the publication of Wittgenstein's *Tractatus Logico-Philosophicus*, the idea of such an order was not yet known as a formulated proposition. Science was (and still is, actually) dominated only by a mechanistic, cause-and-effect approach, which may be considered in its broad sense as a deterministic one too. Wittgenstein moves clearly beyond this paradigm. His *Investigations* are full of metaphors which lack the exactness and orderliness of *Tractatus*. It is a natural consequence of an epistemological decision; a change in the way in which Wittgenstein perceived the world or in which the world seemed to him at that point: random, complex, non-homogeneous, dynamic. This world is manifested in names in an equally diverse, multifaceted way, full of interactions, whose multiple relationships stabilize for a brief while, as senses in some circumstances. He claims that we create notions "as in spinning a thread we twist fibre on fibre. And the strength of the thread does not reside in the fact that some one fibre runs through its whole length, but in the overlapping of many fibres" (Wittgenstein 1958, 32).

Networks

It could be considered a coincidence of notions, but Fritjof Capra referred to the very same notion: life, trying to depict a certain new model describing the structure of the world, especially life as its highly complex and—in this sense—representative form of being (Capra 1997, 2002). The model in question was a network. It is governed by a logic of complexity, which is something much more than just a quality of phenomena; it is a generalized way of understanding of the world, and, with time, also a foundation of a separate branch of science, based on this very way. Capra is one of the many, or even very many authors describing the various contexts of this sort of approach. In 2005, Urry published an article in which he announced a significant turn in research, similar to earlier ones, such as postmodernism, a linguistic turn, or Marxism, meaning one initiating a certain substantial change in the current way of thinking, forcing one to fundamentally revise the adopted views, research tools, or notional apparatus. He called the new turn *the complexity turn* (Urry 2005).

John Urry was a sociologist, with a specific type of approach to the idea of complexity. To him, it is a fact occurring in the social space of science along with scientific employees, scientific institutions, and its broad social setting. This conglomeration can therefore suggest that essentially, the object of observation is the entire social world, where an occurring trend—a new type of outlook on the world—becomes significant and noticeable as well as fashionable and popular. Urry offers a long list of texts, the dates of publishing of which date back to the 1990s, with one item from the late 1980s, and many from the early twenty-first century. The source scientific disciplines of their authors are different, with the most common of them being a social, cultural, or epistemological reflection, but there is also physics, mathematics, or biology. A new, non-simple approach to the understanding of the world and to depicting it in the form of a new model is born. Many people in many places will witness all sorts of mass phenomena and see a need to understand them in a general, synthetic way, so far unknown.

This exact process is described in detail by Capra (2002). Seeing the whole as something different from the sum of its parts, as he claims, appears first as an idea of a system.

Such thinking heads towards itself; it immerses itself in an autotelic problem of its own sense and order. His father, Ludwig von Bertalanffy, formulated his ideas in the 1930s, but only the article published in 1950, entitled *The Theory of Open Systems in Physics and Biology* (von Bertalanffy 1950), earned him some public attention (Skyttner 2005, 39). The year 1954 was when the International Society for General Systems Theory was established, with its main founders being Ludwig von Bertalanffy and Kenneth Boulding. Skyttner stressed that since the very beginning, they had been driven by a sense of pursuit of a different type of science, founded on different grounds, positioned differently in the order of research endeavors. The group of founders, composed of representatives of different disciplines, wanted to go for a universal science:

"They wanted to link together the many splintered disciplines with a law of laws applicable to them all" (Skyttner 2005, 39). Attention to phenomena studied by means of various specific methods typical to particular sciences is not enough anymore. It is science and knowledge—as its appropriate domain—that once again demand to become objects of deep reflection. Eventually, this reflection bears fruit in a distrust of scientific judgment, a growing sense of entanglement in the various extra-scientific contexts: culture, politics, customs, social constraints. Science as an object of research, not as a transparent tool—meta-science—opens the door to new, unusual areas of reflection.

Skyttner considers complexity science also a fruit of systems approach. Castellani and Hafferty argue that it came into being out of the need to create a common and universal idea of a complex system integrating the many various current structures and constructs: "complex adaptive systems, selfreferential systems, autopoietic systems, dynamical systems, systems near chaos, dissipative systems, non-equilibrium open systems, self-organizing systems, emergent systems, and living systems" (Castellani and Hafferty 2010, 119). This universal idea is not related to any specific branch of science, but acts as a synthetic, general approach. It emerges spontaneously in a group of researchers such as: Robert Axelrod, Per Bak, George Cowan, John Holland, Stuart Kauffman, Christopher Langton, Niklas Luhmann, Humberto Maturana, Ilya Prigogine, and Francisco Varela in the 1970s. Castellani offers a map that illustrates the topography of the emerging and developing complexity science: *the complexity map*. It is composed of the ideas and theories along with their authors, depicted in the form of a timeline dating back to the 1940s and stretching to the early twenty-first century (Castellani and Hafferty 2010, 244).[1]

The main property of the points on Castellani's map is their mutual interconnection, expanding dynamically over time. The term "map" can be misleading here, though. This is about both the points on the timeline, which determine the main and separate trends, running from the past to the present, and their positions with respect to each other. These positions are often determined by their mutual relationships. We are dealing rather with a growing network, modifying over time. Growing—because already the earliest researchers whose names are mentioned by Castellani and Hafferty formed a network of collaborators. Over time—the "map" is a process, in fact. A look at the entirety of the events presented therein gives a false sense of stability. It is not, however, a bird's eye view of some territory, but an artificial synthesis of past events. Many years ago, Capra described this structure as a new model appearing in the understanding of the nature of the dynamics of development of

[1] Castellani is still working on his map, with several new versions existing. One of them covers the period up to 2015 (https://sacswebsite.blogspot.com/2013/07/new-version-of-complexity-map.html). It seems that the most recent version dates back to 2018 and can be found at https://www.art-sciencefactory.com/complexity-map_feb09.html (websites accessed on: 82019).

science and knowledge: "the metaphor of knowledge as a building is being replaced by that of the network" (Capra 1997, 39).

Castellani's "map" is a living illustration of this observation; the structure in question is surely a network of knowledge about complexity science. There is one more striking similarity here: it is, after all, a clear image of twisted fibers forming a thread Wittgenstein referred to in a quotation featured earlier. The thread is strong thanks to the fact that its fibers overlap. The connections between the fibers are therefore present in his metaphor, but not yet so clear. Still, there is no doubt: the only time Wittgenstein uses the notion of network in his *Investigations*, it is when he uses the already quoted comparison, just a dozen lines earlier. The thread metaphor illustrates the idea of overlapping and interweaving similarities of names, meaning the idea of the network formed by language (Wittgenstein 1958, 32).

Looking at the world from the point of view of a whole, calling for rising above detailed studies of particular sciences, has brought the problem of knowledge back into the spotlight out of necessity. After all, knowledge is a common platform of all individual efforts. But this way we arrive at a new object of research, found at a different, higher level of reasoning: the synthetic level, and abstract in nature since it is distanced from the level of immediate experience where scientific research usually takes place. This transition is noticed and described by Ackoff in 1972. He defines it as "the difference between the synthetic thinking of a metadiscipline and the analytical thinking of a discipline" (Skyttner 2005, 41). Entering the meta-level, a noteworthy fact, offers a liberation from the said problem of time. A network that develops over time, in a synchronous perspective which is possible when one moves to a higher level of the abstract, is present in its entirety. A funny—but surely not incidental—similarity is that it plays a certain game between the temporary and the stable. This similarity, opening the door to the deepest levels of understanding of time and of its sense, especially in the context of human existence, is associated with an object such as a thread in culture and its manifestations. It has an ancient, Greek, mythological source. It is ingrained deeply and permanently in our symbolic 'storage space' by the image of the Moirai. According to the impact of circulation of some subjects or associations—topoi—codified in literature by Curtius (1948), it cannot be a pure coincidence.

A network, as a universal and common model of processes shaping the world, becomes popular relatively late. Its history as a concept is very old, but as a systematized idea and theory—relatively new. It also develops in an interesting way because it is a story of a conflict that becomes visible when a network analysis of network researchers is conducted. The history of a network as a systematized approach begins as a story of a tool used in social studies. Freeman, one of the most prominent researchers and historians, noticed certain elements he considered evidence of such an approach in the thirteenth century, or even earlier. A full and conscious materialization of the foundation of the approach in question, which can be regarded as the beginnings of the social network theory, emerged only in the 1930s (Freeman 2011, 2004). Zhang slightly shifts this moment in time and mentions here such aces as Durkheim and Tönnies as its precursors. He also sees such an approach in Simmel's output, and—in the 1920s—among British anthropologists and American researchers from Harvard (Zhang 2010).

The social network theory develops eventfully until the 1970s, Freeman recalls, engaging plenty of research centers, with many well-known figures—such as Claude Lévi-Strauss, Robert Merton, Everett Rogers, or Ithiel de Sola Pool—mentioned in this context. In the 1970s, following the efforts of Harrisona White and his students, the social network theory becomes a standard paradigm and a separate field of research (Freeman 2011, 27, 2004, 124). But this organized field and paradigm is affected by a rapid turn in the 1990s, to which Freeman refers bluntly as a revolution. Its authors include several researchers, i.e.: Watts, Strogatz, Barabási, and Albert, who publish two groundbreaking works on the laws governing the network structure at the end of the 1990s (Barabási and Albert 1999; Watts and Strogatz 1998).

These researchers differ from their predecessors because they are physicists, who, in addition, do not take the achievements of the social network theory into consideration: "They simply claimed research topics that had always been part of social network analysis and made them topics in physics" (Freeman 2011, 28). It must have made the explorers of the idea of a network in the field of sociology of the time irritated and even jealous, but it opened the door to other physicists into the world of social networks. This happened for a simple reason. It appeared that one did not have to be competent in the field of social sciences to deal with social networks! They were just networks, governed by universal laws. Anyone trying to enter the domain of the network theory, reviewing the available literature, knows the feeling: a surprising cognitive dissonance between the approach based on social sciences and the approach based on exact sciences. It was easy to get lost in the body of publications from fundamentally different fields but touching upon the same phenomenon. The situation was unusual enough that Freeman analyzed it using statistical data. Both communities: that of physicists and that of sociologists formed two clearly different networks. Interestingly enough, the division survived into later times, at least up to 2015 (Hidalgo 2016).

A network ceased to be only a model of social phenomena. In the second half of the 1990s, the Internet saw a rapid exponential growth, as illustrated by the number of computers connected to the World Wide Web. The World Wide Web is a large network of Uniform Resource Locators. It can be examined as a specific autonomous structure (Albert et al. 1999; Broder et al. 2000) but also as a new social phenomenon (Castells 1996; van Dijk 1999). Its network nature is revealed in a growing number of global phenomena. Physicists are joined by biologists, studying protein networks or metabolic chains. It appears that economic, ecological, technical, communication, physical, and other structures can be effectively presented in the form of networks. Only two–three years after the said groundbreaking articles by himself, Watts, Strogatz, and Albert, Barabási writes as follows: "We have come to see that we live in a small world, where everything is linked to everything else. We are witnessing a revolution in the making as scientists from all different disciplines discover that complexity has a strict architecture. We have come to grasp the importance of networks" (Barabási 2002, 7).

Under the influence of an extraordinary significance of the architecture of networks to science, resulting, on the one hand, from its universal nature that elevates it above the traditional differences between scientific disciplines and, on the other, from its popularity in the world, in 2006, Newman, Barabási, and Watts (physicists by education) call for the establishment of a new science of networks, mentioned first a few years earlier (Newman et al. 2006). They offer several reasons behind their demand, such as the empirical approach or the changeability of networks, but the reason raised as the last is fundamental: "to understand networks not just as topological objects, but also as the framework upon which distributed dynamical systems are built" (Newman et al. 2006, 4).

The notion of a dynamical system has its strict physical interpretation proposed by Poincaré and elaborated on in a classical work by Birkhoff, who is considered the originator of the notion (Abraham et al. 1980; Birkhoff 1966 [1927]; Nolte 2015). An informal summary of the definition in question means a set of interconnected elements, whose parameters change over time. These changes can be described as specific trajectories of the set under consideration in the adopted system of coordinates, built on the basis of the said parameters. The incorporation of mathematical notions and structures—borrowed from the description of a network—into the description of a dynamical system is an innovative solution. But the underlying idea is much deeper: it means associating the network logic with the logic of the most essential, physical description of the world, based on the idea of changeability and movement, which belongs to the most ancient human observations and premonitions regarding how the world exists. And this means the deepest possible level of searches and explorations, a level upon which the interpretations of this existence are founded. The term "deepest" is only a metaphor, though, originating from a certain human habit. A correct interpretation should refer to the highest level, taking the level of the abstract generalization occurring here into account. It is a synthetic view, encompassing the broadest possible—meaning by default: the entire area of the world.

Before we proceed to the last, conclusive subject, emerging as if necessarily in this deep and fundamental area of reasoning—to the idea of complexity, let us recapitulate on the argumentation presented so far. The starting point is two words which can offer a reasonable description of the world: digital transformation. Yet, these two words, belonging to the domain of language, whatever their ethnic or geographic origin may be, they form, somewhat mysteriously, a relationship with the world. This process is discussed at length by Foucault in his most famous work (Foucault 1966). It aims to evidence an idea of a great change taking place in Europe at the turn of the 18th and 19th centuries at the fundamental level because it concerns the foundations of the way in which we describe the world in all kinds of attempts, concerning philosophy, natural history, theory, and pragmatics. Foucault claims that the change involves abandoning the observation of things themselves, of their qualities, which he considers a pretext to look for and formulate all sorts of similarities. We shift our attention as if away from things. This means that in we perceive them as dynamic forms, and this dynamism, this changeability is the way to understand them and position them in the general order. In the introduction,

Foucault writes as follows: "the theory of representation disappears as the universal foundation of all possible orders; language as the spontaneous tabula, the primary grid of things, as an indispensable link between representation and things, is eclipsed in its turn; a profound historicity penetrates into the heart of things, isolates and defines them in their own coherence, imposes upon them the forms of order implied by the continuity of time; the analysis of exchange and money gives way to the study of production, that of the organism takes precedence over the search for taxonomic characteristics, and, above all, language loses its privileged position and becomes, in its turn, a historical form coherent with the density of its own past." (Foucault 2002, xxv).

In the act of describing the world, Foucault sees a need to refer to a thought structure that will take over a role (competence) involving getting to know the world, characterized by a different nature than knowledge, being its effect, or a language acting as a time machine for this knowledge. He calls this structure *episteme*. It is a sphere where an order governing the way in which the world is explored and learned of exists, strongly connected with time and space, which makes it strongly connected to culture as well. It is changeable and diverse, although it features some dominant features materialized as current, general principles of making judgments about the world. For this reason, it is possible to speak of different *epistemes*; Foucault considers the *episteme* of the Western world or culture in two rapid discontinuities: a classical one, which appeared around the seventeenth century, and a contemporary one, emerging at the beginning of the nineteenth century, but also defining the twentieth century.

Reading Foucault's work is homework never fully completed, a lesson in its general and brilliant synthesis, illustrating the dynamics and the nature of the knowledge upon which the idea presented in this book is also based. The idea seems almost obvious in the light of the propositions made after the publishing of Foucault's book in 1966. The author himself could see its beginnings because they were clearly noticeable already before his passing away in 1984. The said propositions concerned synthetic and general models, such as the general theory of complexity, which were based exactly on the idea to put things themselves aside and focus on their structures, meaning certain qualities and relationships surfacing in their coexistence.

At the same time, those propositions tended to break a certain significant barrier of generality, elevating, quite consciously, the analysis and diagnosis of the world to the level of self-knowledge, which seems analogous to that presented by Foucault. The said models of a network or a system are clear attempts of general and superior-tier cognition, formulated by means of different, specific notional structures. These structures belong out of necessity to the languages of exact or natural sciences, different from philosophical, anthropological, or generally humanistic analyses. The analogy between Foucault's *episteme* and models such as a system or a network is quite distant. One could dispute over their analytical parallelism given the lack of a good, faithful translation between the languages they use, but it is hard not to notice the possibility Foucault had clearly anticipated. It is possible to achieve and utilize exactly such a high-tier level of knowledge.

The models in question are a direct and purposeful materialization of a certain meta-level of knowledge. From the point of view of this meta-level, the tools enabling one to engage in analytical activities on a daily basis, to think, reason, speculate, are used *ad hoc*, selected and picked according to the set objective. Like a hammer, suited to driving nails into walls much better than a particle accelerator, although the latter is not left out as an alternative option entirely. This is the very same way in which we use language, maintaining a close relationship with it: forced, but bearing in mind that it makes a selected, more or less expected, more or less appropriate modus of cognition reality. (Using a hammer may also involve some trouble). And this is how the apparently banal phrase of "digital transformation" should be perceived as well.

Complexity

In the passage quoted earlier, Barabási equates network with complexity. A network is a practical materialization of complexity in the aspect that concerns the relational order of many interconnected elements. A network is a morphology of complex systems, which have many other properties apart from being complex. In 2002, the same year in which Barabási's work is published, Capra's book sees the light of day. Its main theme is life, which, as the author claims, can be described in synthetic terms: "A central insight of this unified, systemic understanding of life is that its basic pattern of organization is the network. At all levels of life-from the metabolic networks inside cells to the food webs of ecosystems and the networks of communications in human societies the components of living systems are interlinked in network fashion" (Capra 2002, 261).

Capra formulates the idea in a brief manner as follows: "wherever we see life, we see networks" (Capra 2002, 6). In the introduction, he makes mathematical theory, "a body of mathematical concepts and techniques", the complexity theory or, speaking more precisely, non-linear dynamics the foundation of the analysis of life in the broad sense (Capra 2002, xviii).

Almost exactly half a century after the publishing of Wittgenstein's *Philosophical Investigations*, Capra defines life in a fashion favorable to the argumentation presented here. His books cited herein are not scientific monographs per se; they are rather extensive essays, the competence and vision of which cannot be denied. He uses the notion of life in a very broad sense and encompasses both biological and social dimensions therein. The latter is highlighted by its communicative aspect, which is clearly understood as fundamental, essential, providing the foundation for other social relationships. A broader depiction than that of Wittgenstein, but quite similar. Wittgenstein seems to see life as a certain space of events. A specific one as it is limited to those events which become somehow triggered by human activity related to language, also communicative activity *in re*. Wittgenstein argues: "to imagine a language means to imagine a form of life" (Wittgenstein 1958, 8). The similarity between Wittgenstein and Capra is interesting and justified even though life as viewed by Wittgenstein remains mysterious and spacious, and his vision is more of a premonition, an image. In the case of Capra, life is more concrete, taking the form of a complex network structure, which itself is a disciplined theory.

The intuition of Wittgenstein, who might have been familiar with the basics of the network theory, finds evidence to the fullest half a century later as a concrete, complex architecture. Interwoven notions, entwined additionally in the circumstances of their applications, become tangled in life in a twofold manner: firstly, they emerge as a result of their own revelation, which takes place, after all, only through interaction and relationships with other notions. Secondly, these notions involve instances of concretization that have already occurred, forming vast storehouses of possibilities since words mean something, even though this "something" is not unambiguous. Such a vision becomes a starting point for many explorations of significance to the humanities and social sciences. A colossal space of the understanding of the dynamics of language, knowledge, and their social and cultural circumstances opens. The consequences of these reflections are far-reaching; they enter the area of politics, reconstruct the basic philosophical notions, and reach the very heart of their own sciences: to the construct of human. Such figures as Michel Foucault, Jean-François Lyotard, and many others pave the way to the reflection that has continued till this day.

Let me recall here two philosophers who have thoroughly studied the described process of the existence of language: John Longshaw Austin and Jacques Derrida, although they have pursued their studies in different analytical conditions and within the framework of a different conceptual discipline. The former included colloquial language in the area of reflection. A language which is the richest, alive, changeable, and, in principle, incorrect, but effective. The name the others referred to the procedure of deciphering the complex nature of language and the world at the same time as deconstruction, which he described as follows: "Deconstruction does not consist in passing from one concept to another, but in overturning and displacing a conceptual order, as well as the nonconceptual order with which the conceptual order is articulated." (Derrida 1984, 329). It is clear that clusters of words so perceived—a digital transformation so perceived—repeat the dynamics of life within their own dynamics, but irreversibly interwoven with life if life is considered as the "nonconceptual order" to which Derrida refers. In the said order, we have discovered an architecture we would call a network architecture by today's standards. The architecture in question points to a paradigmatically different way of perception of this order, which seems to fulfill the conditions of complexity. Earlier, we discussed, to some extent, the historical circumstances of the emergence of this idea as a strict, disciplined, scientific approach: the science of complexity. The description of the theoretical basis of this science, which now seems to be necessary, is at the same time easy and difficult for the same reason: a vast sphere of reflection.

The description can begin from a book by Melanie Mitchell. A book for the general public, but one highly insightful and reliable (Mitchell 2011). She appears to have been successful in her mission since the book was awarded the Phi Beta Kappa Science Book Award in 2010. Melanie Mitchell is affiliated with the famous Santa Fe Institute, established in 1984, with its claim featured on its website reading as follows: "Searching for Order in the Complexity of Evolving Worlds" (Santa Fe Institute 2019). Mitchell's definition of complexity is the following: "a system in which large networks of components with no central control and simple rules of operation give rise to complex collective behavior, sophisticated information processing, and adaptation via learning or evolution" (Mitchell 2011, 13). A bit further on, Mitchell adds a functional supplementation, referring to it as "a system that exhibits nontrivial emergent and self-organizing behaviors". The strategies of describing a complex system often address a list of its features. Different authors create such lists in a slightly different manner, and a complex system's vision we can arrive at oscillates between the sum or the shared parts (product) of such lists.

Here they are.

The relatively oldest list, proposed by Ciliers and Brucea:

- (i) Complex systems consist of a large number of elements;
- (ii) The elements have to interact, and this interaction must be dynamic. A complex system changes with time. The interactions do not have to be physical; they can also be thought of as the transference of information;
- (iii) The interaction is fairly rich;
- (iv) The interactions themselves have a number of important characteristics. Firstly, the interactions are non-linear; the interactions usually have a fairly short range"
- (vi) There are loops in the interactions;
- (vii) Complex systems are usually open systems;
- (vii) Complex systems operate under conditions far from equilibrium;
- (ix) Complex systems have a history;
- (x) Each element in the system is ignorant of the behaviour of the system as a whole, it responds only to information that is available to it locally (Cilliers and Bruce 1998, 3).

Johnson's descriptive list:

- The system contains a collection of many interacting objects or "agents".
- These objects' behavior is affected by memory or "feedback".
- The objects can adapt their strategies according to their history.
- The system is typically "open".
- The system appears to be "alive".
- The system exhibits emergent phenomena which are generally surprising, and may be extreme.
- The emergent phenomena typically arise in the absence of any sort of "invisible hand" or central controller.
- The system shows a complicated mix of ordered and disordered behavior (Johnson 2007, 13 passim).

A succinct list as offered by Ladyman et al., limited to fundamental features:

- Nonlinearity.
- Feedback.

- Spontaneous order.
- Robustness and lack of central control, emergence.
- Hierarchical organization.
- Numerosity (Ladyman et al. 2013, 35 passim).

A similar list drawn up by Byrne and Callagan, including phenomena associated with complex systems:

- Nonlinearity.
- Emergence.
- Far from equilibric systems.
- Phase space and attractors.
- Autopoiesis, boundaries, structure and hierarchies (in complex systems).
- Control parameters ("changes in those elements can have an effect such that the nature of the system is changed in a qualitative fashion") (Byrne and Callaghan 2014, 17 passim).

A list proposed by Holland, member of the renowned Santa Fe Institute, a legendary institution studying complex systems with the aim to make the findings accessible to the general public:

- self-organization into patterns, as occurs with flocks of birds or schools of fish;
- chaotic behaviour where small changes in initial conditions ('the flapping of a butterfly's wings in Argentina') produce large later changes ('a hurricane in the Caribbean');
- 'fat-tailed' behaviour, where rare events (e.g. mass extinctions and market crashes) occur much more often than would be predicted by a normal (bell-curve) distribution;
- adaptive interaction, where interacting agents (as in markets or the Prisoner's Dilemma) modify their strategies in diverse ways as experience accumulates.
- (…) In addition, as already mentioned, emergent behaviour is an essential requirement for calling a system 'complex'. (Holland 2014, 5).

The presented lists are similar and differ, basically, in the kind of approach, which may be either more exact or more descriptive, aspiring to be either precise or simple. All of them elaborate more or less on Mitchell's definition, supplementing it or making it more specific. The feature they all share is assuming a detached research position, which—by definition—breaks away from the level of narrow isolated systems, aiming at a more general perspective. Naturally, the number of elements of an observed system—and the diversity of these elements—grows with the number and the kind of relationships existing between them, reaching a scale unavailable to traditional cause-and-effect models. There appears a need for a phasic, paradigmatic change of the observation procedure because, as underlined many times, complex systems reveal non-standard behavior, one beyond predictions based on fragmentary interactions of elements. This leads to emergence or autopoiesis, meaning organized, phenomenal effects observable only at the level of a system as new, unexpected, etc. By assumption, such an effect is digital transformation, and the environment where it takes place, the complex system that lays the foundation for it to happen, is knowledge, articulated in language.

References

Abraham, Ralph, Jerrold E. Marsden, Tudor Ratiu, and Richard Cushman. 1980. *Foundations of mechanics: 2nd edition*. 2nd ed. Reading: Addison-Wesley Publishing Company.

Albert, Réka, Hawoong Jeong, and Albert-László Barabási. 1999. Diameter of the world-wide web. *Nature* 401 (September): 130.

Armstrong, David Malet. 1997. *A world of states of affairs*, Cambridge studies in philosophy. Cambridge: Cambridge University Press.

Barabási, Albert-László. 2002. *Linked. The new science of networks*. Cambridge, MA: Perseus Pub. http://www.ebrary.com.

Barabási, Albert-László, and Réka Albert. 1999. Emergence of scaling in random networks. *Science* 286 (5439): 509. https://doi.org/10.1126/science.286.5439.509.

Birkhoff, George D. 1966. *Dynamical systems [American Mathematical Society colloquium publications, volume 9]*. Revised edition: American Mathematical Society.

Broder, Andrei, Ravi Kumar, Farzin Maghoul, Prabhakar Raghavan, Sridhar Rajagopalan, Raymie Stata, Andrew Tomkins, and Janet Wiener. 2000. Graph structure in the web. *Computer Networks* 33 (1): 309–320. https://doi.org/10.1016/S1389-1286(00)00083-9.

Burgin, Mark. 2015. *Theory of knowledge: Structures and processes*, World scientific series in information studies: Volume 5. Vol. 5. Singapore: World Scientifc Publishing Co. Pte. Ltd.

Byrne, David S., and Gill Callaghan. 2014. *Complexity theory and the social sciences: The state of the art*. New York: Routledge, Taylor & Francis Group.

Capra, Fritjof. 1997. *The web of life: A new scientific understanding of living systems*. New York/London: Anchor.

———. 2002. *The hidden connections: Integrating the biological, cognitive, and social dimensions of life into a science of substainability*. 1st ed. New York: Doubleday.

Castellani, Brian, and Frederic William Hafferty. 2010. *Sociology and complexity science: A new field of inquiry*, Softcover reprint of hardcover 1st ed. 2009th ed. Berlin: Springer.

Castells, Manuel. 1996. *The information age: Economy, society, and culture. Vol 1: The rise of network society. Vol. 2: The power of identity. Vol. 3: End of the millennium*. Oxford: Blackwell.

Chappell, Timothy D.J. 2005. *Reading Plato's Theaetetus*. Indianapolis/Cambridge: Hackett Publishing Company Ltd.

Cilliers, Paul, and Tony Bruce. 1998. *Complexity and postmodernism: Understanding complex systems*. London: Routledge.

Curtius, Ernst Robert. 1948. *Europäische Literatur und lateinisches Mittelalter*. Bern: A. Francke.

Dancy, Jonathan, Ernest Sosa, and Matthias Steup, eds. 2010. *A companion to epistemology*. Malden: Wiley-Blackwell.

Derrida, Jacques. 1984. *Margins of philosophy*. Trans. Alan Bass. Reprint ed. Chicago: University of Chicago Press.

Deschoux, M. 1980. *Platon Ou Le Jeu Philosophique*, Annales Littéraires de l'université de Franche-Comté. Paris: Presses universitaires de Franche-Comté, Le Belles Lettres. https://books.google.fr/books?id=na1ZzQTI_MYC.

Fleck, Ludwik. 1935. *Entstehung und Entwicklung einer wissenschaftlichen Tatsache: Einführung in die Lehre vom Denkstil und Denkkollektiv*. Basel: Benno Schwabe.

Foucault, Michel. 1966. *Les mots et les choses. Une archéologie des sciences humaines*. Paris: Gallimard.

———. 2002. *The order of things: An archaeology of the human sciences*. London: Routledge.

Freeman, Linton C. 2004. *The development of social network analysis: A study in the sociology of science*. Vancouver/North Charleston: Empirical Press.

———. 2011. The development of social network analysis – With an emphasis on recent event. In *The Sage handbook of social networks analysis*, ed. John Scott and Peter J. Carrington, 26–39. London: SAGE Publications.

Gödel, Kurt. 1931. Über formal unentscheidbare Sätze der Principia Mathematica und verwandter Systeme I. *Monatshefte für Mathematik und Physik* 38 (1): 173–198. https://doi.org/10.1007/BF01700692.

Hidalgo, César A. 2016. Disconnected, fragmented, or united? A trans-disciplinary review of network science. *Applied Network Science* 1 (1): 6. https://doi.org/10.1007/s41109-016-0010-3.

Holland, John H. 2014. *Complexity: A very short introduction*. Oxford: Oxford University Press. http://public.eblib.com/choice/publicfullrecord.aspx?p=1696424.

Johnson, Neil F. 2007. *Simply complexity: A clear guide to complexity theory*. Oxford: Oneworld Publications.

Klement, Kevin. 2016. Russell's logical atomism. In *The Stanford encyclopedia of philosophy*, ed. Edward N. Zalta. Metaphysics Research Lab, Stanford University. https://plato.stanford.edu/archives/spr2016/entries/logical-atomism/.

Ladyman, James, James Lambert, and Karoline Wiesner. 2013. What is a complex system? *European Journal for Philosophy of Science* 3 (1): 33–67.

Maciag, Rafal. 2022. Theory of knowledge based on the idea of the discursive space. *Philosophies* 7 (4): 72. https://doi.org/10.3390/philosophies7040072.

Minsky, Marvin. 1974. *A framework for representing knowledge*. Cambridge, MA: Massachusetts Institute of Technology.

Mitchell, Melanie. 2011. *Complexity: A guided tour*. 1st ed. Oxford: Oxford University Press.

Nagel, Ernest, and James R. Newman. 1958. *Gödel's proof*. New York: New York University Press.

Newman, Mark, Albert-László Barabási, and Duncan J. Watts. 2006. *The structure and dynamics of networks*. 1st ed. Princeton: Princeton University Press.

Nolte, David D. 2015. *Introduction to modern dynamics: Chaos, networks, space and time*. 1st ed. Oxford/New York: Oxford University Press.

Plato. 1856. *Platons Werke*. Trans. Friedrich Schleiermacher. Berlin: Georg Reimer. http://archive.org/details/platonswerke00schlgoog.

———. 1987. *Theaetetus*. Trans. Robin H. Waterfield. Harmondsworth/New York: Penguin Classics.

Platon. 2002. *Parmenides, Teajtet. Przełożył oraz wstępami i objaśnieniami opatrzył Władysław Witwicki*. Kęty: Antyk.

Quine, Wilard V. 1957. The scope and language of science. *The British Journal for the Philosophy of Science* VIII (29): 1–17. https://doi.org/10.1093/bjps/VIII.29.1.

Quine, W.V. 1995. *From Stimulus to Science*. Harvard University Press, Cambridge, Mass.

Russell, Bertrand. 1948. *Human knowledge – Its scope and limits*. London: George Allen and Unwin Ltd.. http://archive.org/details/in.ernet.dli.2015.461169.

———. 2002. Introduction. In *Tractatus Logico-Philosophicus*, ed. Ludwig Wittgenstein, 2nd ed., ix–xxv. London/New York: Routledge.

———. 2003. Analytic realism. In *Russell on metaphysics. Selections from the writings of Bertrand Russell*, ed. Stephen Mumford, 1st ed., 91–96. London: Routledge.

Santa Fe Institute. 2019. About | Santa Fe Institute. 2019. https://www.santafe.edu/about.

Skyttner, Lars. 2005. *General systems theory: Problems, perspectives, practice*. 2nd ed. Hackensack: World Scientific. http://search.ebscohost.com/login.aspx?direct=true&db=nlebk&AN=210756&site=eds-live.

Urry, John. 2005. The complexity turn. *Theory, Culture & Society* 22 (5): 1–14.

van Dijk, Jan. 1999. *The network society*. London: SAGE Publications.

Van Harmelen, Frank, Vladimir Lifschitz, and Bruce Porter, eds. 2008. *Handbook of knowledge representation*. Amsterdam/Boston: Elsevier.

von Bertalanffy, Ludwig. 1950. The theory of open systems in physics and biology. *Science* 111 (2872): 23. https://doi.org/10.1126/science.111.2872.23.

Watts, Duncan J., and Steven H. Strogatz. 1998. Collective dynamics of 'small-world' networks. *Nature* 393 (6684): 440–442. https://doi.org/10.1038/30918.

Weller, Kathrin. 2010. *Knowledge representation in the social semantic web*. Berlin/New York: De Gruyter Saur.

Wittgenstein, Ludwig. 1958. *Philosophical investigations*. 2nd ed. Oxford: Basil Blackwell.

———. 1974. *Tractatus Logico-Philosophicus*. 2nd ed. London/New York: Routledge.

Wolniewicz, Bogusław. 2000. O Traktacie. In *Tractatus logico-philosophicus*, ed. Ludwig Wittgenstein and Biblioteka Klasyków Filozofii. Warszawa: Wydawnictwo Naukowe PWN.

Zhang, Mingxin. 2010. Social network analysis: History, concepts, and research. In *Handbook of social network technologies and applications*, ed. Borko Furht, 3–21. Boston: Springer US. https://doi.org/10.1007/978-1-4419-7142-5_1.

Chapter 4
Gettier and the Problem of Language

The year 1963 saw one of the most interesting episodes of man's struggle with the world around. A struggle where the mind proved to be still most useful. A young philosopher by the name of Edmund Gettier, as described faithfully by Pritchard (2006, 26), was worried about his employment at the university. It depended on the number of publications—of which he had too few at the time. This coincidence already reveals the rather banal irony of a philosopher's fate, who is subject to the mundane laws of existence like everyone else. But as the plot developed, Gettier was apparently guided by an inadvertent sense of a relevant message, for in browsing through the possible subjects for the necessary article, he chose epistemology, which seemed interesting to him by promising quick results in the form of imminent print and thus enabling him to complete the task soon. Epistemology is the science of knowledge, so it is also the story of the jungle in which those who seek this knowledge professionally live. Gettier had not been particularly interested in it before, but this lack of interest did not prove significant. Yet at that point, he found himself in a situation where it suddenly appeared very close to him. He dealt with a specific 'species' of knowledge—its most obvious and therefore, quite understandably, closest to experience form. The so-called propositional form—a form based on propositions. In the aforementioned year 1963, Gettier publishes a short, three-page text which causes a real upheaval (Gettier 1963). However, he himself does not engage in the discussion he provoked. He also abandons epistemology and writes nothing more about it later on. The story ended really well. Gettier retained his position because his text was published in a popular journal.

All the steps that the young philosopher took testify to his far-reaching pragmatism and knowledgeability. Assuming that knowledge is also one of the gifts that philosophers bestow upon us, it appears that Gettier did not take advantage of the opportunity to share neither knowledge nor any more of its assets, revealing them only to a very slight extent. But this superficial judgment fades and vanishes when the content and subject matter of his text are weighed, and the legendary and standard repertoire of anecdotes about him gains in complexity and depth. The article by Gettier openly and loudly objected to the definition of knowledge that had prevailed

R. Maciag, *Knowledge as a Tale*, https://doi.org/10.1007/978-3-031-69820-0_4

since Plato's time, and its author simply and succinctly demonstrated the fallacy in its construction. This fallacy is that it is possible to satisfy all of the preconditions for knowledge and yet not have knowledge, which seems paradoxical because the impression that there is a situation where we simultaneously 'know' and 'don't know' occurs. The reality of knowledge, however, turns out to be much more complex, and Gettier's insights lead to far-reaching conclusions that are particularly relevant to this text. His observations are part of a whole series of events that significantly change the way knowledge reveals itself to us.

In his text, Gettier cites Plato's *Theaetetus* and *Menon*. Plato's *Dialogues*, which the said texts are part of, however, are not separate texts, alien to and independent of each other. Instead, they form a certain complete universe. This universe sees a variety of threads and themes, arranged in the shape of a current configuration that supports the theme being analyzed, interweaving with each other. This is also the case with the thread concerning knowledge as described by Cornford (1935). This thread, together with accompanying and approximate additions, also appears in texts such as *Phaedo*, *The Symposium*, *Republic*, and *Timaeus*, as recalled by Burgin (2015). The list can actually also include *Sophist* and *The Statesman*. Undoubtedly, however, the most important text offering a concise definition of knowledge is *The Theaetetus*, in which the title character keeps on repeating this definition to Socrates as he does not remember having heard it from someone. These unusual circumstances, however, do not make it any less relevant or valid. It seems reasonable to cite it as it is expressed in a neat four-word formula: *Dóksa alethés metá lógu* aforementioned earlier. The English version found most commonly in the literature reads as follows: "true belief accompanied by a rational account" (Plato 1987, 115), is usually shortened to three letters: JTB, which stand for *Justified True Belief.*[1] It was this definition that came under Gettier's criticism. It consists of three conditions which lead to the state of knowledge if met. This threefold nature must have been impressive, sustaining a mythical sense of completeness until Gettier attacked it.

The article in which Gettier addressed the Platonic definition of knowledge in a critical manner was shamelessly short and mostly consisted of two examples whose central figures had legitimate beliefs consistent with the truth and yet lacked knowledge. The disappointment associated with the failure of our justified belief that turns out to be false is a common experience we all happen to share. We almost constantly justify the judgments we make about the weather or politics, and yet they refuse to come true. Completely non-trivial situations in which such a justified belief is true—but is not knowledge—are an entirely different class of events and require specifically constructed circumstances. Both of Gettier's examples involve a combination of two separate and random facts to construct certain sentences, subject to further logical analysis. In the first example, the central figure is Mr. Smith, who is convinced (this is a legitimate judgment) that he will not get the job (his boss told him so) which is to be given to Jones. In addition, Smith counted the coins in Jones' pocket and arrived at a result of ten. Hence, Smith claims that a man who has ten

[1] Here, we will use the terms of judgment (in the logical sense) and belief interchangeably.

coins in his pocket will get the job. Meanwhile, it turns out that the job was given to Smith. Also, when reached into his pocket, he found ten coins in it too. The conclusion is that Smith was guided in his claim by a reasonable and true belief (everything in his claim was correct: the man who had ten coins got the job), and yet he did not know what he included in that claim; after all, he was thinking about Jones—not about himself.

Example two is more difficult, though similar. Smith has an incorrect but reasonable belief that Jones has a Ford car (Jones invited him for a ride). Smith has another friend—one named Brown. Smith expresses his beliefs in a sentence claiming that Jones has a Ford or Brown is in Barcelona. The certain artificiality of this sentence does not flaw its logic: it is an alternative. And Smith is right, which logically means that it is a true sentence because at least one of the partial sentences is indeed true. This condition, which is also in line with common sense (after all, the sentence "the Sun revolves around the Earth or the Earth revolves around the Sun" is true), is sufficient. However, it is not fulfilled in the situation which the example describes—contrary to what Smith thinks, because Jones does not have a Ford, but instead Brown is indeed in Barcelona. Smith therefore doesn't know what the world is like in the small part that encompasses Jones and Brown and the situations they are in. Thus, while he has a reasonable and true (logical) belief, he still doesn't have knowledge.

Sober supplements Gettier's examples with Russell's his own proposals (Sober 2005, 155). Russell invokes a story about a clock that does not work although its indications are usually on time. The observer doesn't know that the clock doesn't show the time and looks at it at the moment when, by chance, the stopped clock shows the correct time (a clock that has a regular twelve-hour face shows the right time twice a day). Such an observer has thus a reasonable and true belief that the clock is showing the correct time. Yet the reality is different, and the knowledge of this reality is not for the observer to have. Sober's example is even simpler. It's a story about a lottery ticket. The ticket's owner has a reasonable—highly probable—belief that their ticket will not win, and this is exactly what happens in the following week's draw. However, they could not have had such knowledge when they developed this reasonable and ultimately true belief.

Sober argues that in these examples, the subject has "highly reliable but not infallible" evidence for the claim they believe in. That's because they base their belief on what usually happens: usually the boss knows what they're doing, the clock shows the right time, and lottery tickets don't win. This imperfection is part of our perception of the world. It produces effects far deeper and more fundamental than what one can make out of logical games at first glance. It leads to the conclusion that we know nothing at all since our judgments about the world are always based on the uncertain and misleading testimony of our senses. Another thing to add here is that we also know that this imperfection goes deeper, into the realm of our cognitive mechanisms, which has been a problem at least since Freud. But it also reaches higher up, into the realm of epistemic mechanisms that emerge at the social level, which we will consider more extensively here later on, referring to the most important contributor to this realm in the twentieth century—the one and only Michel Foucault.[2]

It is possible to come across traces of such a serious analytical background even before Gettier's era. The issue is addressed by Chisholm, who refers to two examples: Bertrand Russell (already mentioned) and Alexius Meinong. In 1906, the latter wrote about an instrument called the Aeolian harp. This instrument, known since ancient times, produces subtle sounds when a breeze blows over its delicate strings stretched in the air. Remaining within their range can lead to an experience of auditory hallucinations when the strings are actually silent. This example provides excellent support for Sobers' intuition. Russell, on the other hand, wrote as early as in 1948, offering his example of the clock, and another one, concerning misremembered names, that "it is very easy to give examples of true beliefs that are not knowledge", adding that knowledge is a subclass of these beliefs: "every case of knowledge is a case of true belief, but not vice versa" (Russell 1948, 170).

However, it was Gettier's article that started a lively discussion which has so far failed to find a conclusion. He 'forced' his way into epistemology and caused a huge stir, named the Gettier problem as a result. This stir has engaged many philosophers and led to the proposal of a number of various remedies. Many, as Moser writes, have sought a fourth condition to remedy the flaw of Plato's definition, or have sought to strengthen the justification condition so that it becomes unquestionable (Moser 2010). The discussion that Gettier started is covered in detail by Shope (2004). It abounds with names and proposals. Borges et al. even claim that the latter outnumber the former (Borges et al. 2018). The density of this discussion, the universality, and ultimately the lack of a jointly and broadly accepted solution, leads to an irreversible skepticism regarding the Platonic definition and, more profoundly, a quite basic anxiety concerning its very possibility. Sober even argues as follows: "skeptical reasoning denies a fundamental part of our common-sense view describing our relationship with the world. According to common sense, people have knowledge about the world they inhabit; skeptical reasoning suggests that common sense is wrong" (Sober 2005, 157).

It seems reasonable, however, to ask a question about why did we wait as long as until the twentieth century—which is over two thousand years—to arrive at such a sad conclusion? Or what twentieth-century event made such a conclusion come to light? It cannot have been the invention of the Ford automobile, the development of a peculiar socio-economic practice of employment, or the emergence of easy global travel, can it? The concern about the impact of our flawed perceptual tools on what we know about the world, however, is not new. It is necessary to mention here e.g. the beginning of the seventeenth century, when this anxiety became the subject not only of philosophy, i.e. of thinkers such as René Descartes or Francis Bacon, but

[2] The problem that arises in applying the premise of a 'justified belief' is well known in the domain of epistemology also because it leads to the problem of epistemic regress, the object of which is the situation of an infinite chain of justified beliefs when successive such beliefs justify the previous ones. Sexstus Empiricus is cited as one of the first authors to draw attention to this problem (Rescher 2003, 22), although its sources can be found even in Aristotle's writings (Audi 1999, 272).

also of writers such as Pedro Calderón de la Barca or William Shakespeare. They all expressed distrust of the senses, viewing them as carriers of a false perception of the world that could easily be mistaken for e.g. a dreamy illusion or a theatrical performance. It's therefore not the matter of belief where the source of said concern lies. Rather, the dread is caused by the second condition of Plato's definition of knowledge: truth. For it is possible, as Gettier remarked, to have no knowledge and not violate the truth condition at the same time. One can make true judgments that are not knowledge because according to the principle Russell wrote about, knowledge is a subclass of these judgments and is contained in them.[3]

The way out that opens up in this unpleasant situation is not particularly simple, and its convoluted nature does not make this argument any easier. We can skip it and go straight to the conclusions that emerge from it ten paragraphs or so further on because it's not the discussion of Gettier's idea that is at issue, but the tacit assumption that originates in it. According to the intention of this text, it has a dramatic impact on the fate of knowledge as a concept that denotes an idea of a certain phenomenon. It is a sign of a changing attitude towards it, which will dissipate in various ways and eventually lead to its theory—which is the subject of the present discussion.

Let us first look at Russell's example: the observer is correct in claiming that it is nine o'clock when they look at the clock which has stopped at nine o'clock. This clock indicates nine o'clock and it is nine o'clock. An exaggeration here is a generalization implying that the clock works.[4] The mistake is that some time-specific observation is extended over a time of indeterminate but significant length. Time also plays a fundamental role in the case of Sober's example with the lottery ticket: it must be neglected for the example to make sense. The buyer is a pessimist and believes that their ticket has no chance of winning already at the time of buying it, which is pretty well justified based on a very simple probability calculus (there are a thousand tickets and only one wins). This buyer already knows this at the time of purchase and is right, although their belief doesn't materialize as truth until a week later, when the drawing takes place. The generalization here is made at the cost of excluding the time dimension from reasoning. In the first example, its central figure knows wrongly, even though they hold a true and reasonable belief (they would have to watch the clock for a longer time); in the second example, the central figure knows rightly, even though they have no such right (they would have to know the future).

Gettier's examples are more basic. In this case we are also dealing with a generalization of a slightly different nature, one broader or deeper, because it also includes examples offered by other authors. This time, it takes the shape of an initial, tacit,

[3] The twentieth century also brings about a great collapse of the concept of truth, which happens in the realm between philosophy and social sciences, which is described in Zygmunt Bauman's essay titled *On Truth, Fiction and Uncertainty* (Bauman 1997, 112).

[4] I think that a similar intention guides Sober when he cites a reasoning process that refers to what 'usually' happens. It is meant to draw our attention to the fact that we use generalizations when we judge reality.

and fundamental assumption where it is claimed that there is an instance capable of determining truth and knowledge. This instance has two materializations. First, it is a certain logic: it is built upon a set of fundamental rules which assign, under appropriate conditions, a truth value to the conjunction (which is the basis of the example with the boss deciding on who to give the job to and the coins in the pocket) and the alternative (which is the basis of the example with the Ford car and Barcelona). Second, this instance is omniscient by design. Let us leave the first case—the existence of adequate logic—aside, although it is a door to an extremely interesting, widely discussed, and unusual proposition-rich field of logics far beyond the realm of ordinary common sense. Two works by Ludwig Wittgenstein are also examples of this logical grapple with language: *The Tractatus Logico-Philosophicus* and *Philosophical Investigations*. The development of the line of thought of this most important twentieth-century philosopher of language is mainly a story of the justified departure from the grand scheme of describing language in terms of strict logic in favor of its complex and non-obvious nature. This reminder must cause a shiver of anticipation in light of the role the aforementioned works played in shaping the views of language and its role in the twentieth century—a time of great reconstructions and discoveries in this area.

The perspectives of reasoning, brought to being by the second materialization of this mysterious instance that seems to know everything, are no less fascinating, even though they are actually quite simple when one asks the question about the source of knowledge. This question is: who knows all that is necessary to verify the knowledge of the central figure of each of the examples given by Gettier (as well as others)? The question is fully legitimate and correct in the context of the initial definition, which, after all, directly addresses some subject (human) who holds some judgment (belief) about the world. In other words, there must always be such an entity in order to speak of knowledge. According to Gettier's reasoning, such an instance that must be beyond all locality—both temporal and spatial exists, for it is possible to determine if the characters of stories do have knowledge only from the perspective of this very instance. Of course, this perspective and position is realized, for example, by us, the readers of the examples, and their author, having a very general, timeless, universal insight into their reality (though that doesn't mean we know everything, but that we know more, much more, and sort of differently). There is no doubt that we thereby assume an extraordinary position that exists on another level of a story—the meta-level (a story about a story).

This principle extends to include other authors' examples as well. We know, for instance, that the clock in Russell's example doesn't work, which the central figure of the story doesn't know. We also know that Mr. Brown is in Barcelona, though Smith has no idea of it. We're able to compress time because only then, in a timeless world, can we say, following Sober, that all the conditions of Plato's definition were fulfilled on the day the lottery ticket was bought, while in reality—the reality of the lottery ticket buyer—this will happen only in a week's time (on the day of the draw it turns out that the ticket lost—just as its buyer thought the week before). A similar timing trick is required to make Gettier's example about the boss who gives jobs and the ten coins effective (Smith learns about the boss' decision and the coins later than he pronounces his judgment). This metatextual positioning of the knower (the omniscient?) means that although knowledge, as defined, is at the disposal of the central figure of a given story (example), it also exists in another, broader resource, and thus, looking from the perspective of that central figure, also in another, extrinsically existing (non-human?), transcendent way. After all, in the reality in which Smith, Brown, the ticket-buyer, and the clock-watcher find themselves, there is no author or listener to the stories about them, without whom a justification for the imperfection of the Platonic definition of their knowledge would not be possible. It is this conclusion that triggers a real and profound disruption in the previous understanding thereof.

The circumstances we describe here are momentous not only because they have sparked an intense and non-conclusive debate among epistemologists. They seem analogous to another, less commonly known situation faced a little earlier (30 years earlier, to be exact) by a similarly fundamental conceptual construct of truth. The description of this situation, which also takes place in the reality of language, was offered by Alfred Tarski in the 1930s—together with the analysis thereof and the conclusions drawn therefrom. He argued for defining truth as a linguistic phenomenon, or more precisely, a phenomenon related to that realization thereof which involves the practice of the existence of language: an utterance expressed in sentences, which can be treated as a universal event without regard to its physical realization. Tarski bases his strategy on the need to differentiate between different levels of speech, which he describes exhaustively in his famous 1933 text entitled *Pojęcie prawdy w językach nauk dedukcyjnych* (*The Concept of Truth in Formalized Languages*) (Tarski 1933). He distinguishes truth as a certain judgment as opposed to the procedure of defining and verifying it on the basis of a created definition, although both activities always take place on the same plane—which is language. Tarski refers directly to sentences as the area of realization of this language. Many years later, in a text recapitulating his thoughts, published in 1969 in the *Scientific American* journal, he summarizes the misleading and instructive properties of everyday language, for which it is not only impossible to define the concept of truth, but even to use "this concept in conformity with the laws of logic." This situation leads to paradoxes:

> [colloquial language] is *semantically universal* in the following sense. Together with the linguistic objects, such as sentences and terms, which are components of this language, names of these objects are also included in the language (as we know, names of expressions can be obtained by putting the expressions in quotes); in addition, the language contains semantic terms such as "truth," "name," "designation," which directly or indirectly refer to the relationship between linguistic objects and what is expressed by them. Consequently, for every sentence formulated in the common language, we can form in the same language another sentence to the effect that the first sentence is true or that it is false. Using an additional "trick" we can even construct in the language what is sometimes called a self-referential sentence, that is, a sentence S which asserts the fact that S is true or that it is false (Tarski 1969, 67).

In the last words of the quoted passage, Tarski refers to the famous liar paradox, which he covers at length in his text—ultimately resolving it. The source of this paradox is precisely the colloquial language, which mixes two different levels of description: the level of direct judgments about things with the level on which the conditions of correctness and truthfulness of these judgments are defined and the extent of fulfillment of these conditions—such as truth—is evaluated. The paradox is possible to be removed and the concept of truth can be defined at a point where the two levels are separated, creating two different languages—the object language and the metalanguage, to use Tarski's nomenclature. The object language is used to enunciate the truth, while the metalanguage is used to construct its definition and test its 'performance' in practice.

In the case of the examples of Gettier and others described earlier, an educational analogy to the situation that Tarski found and remedied occurs. These examples are expansions of the Platonic definition of knowledge, made using everyday language—and they tend to suffer from an internal contradiction: they simultaneously meet and do not meet the conditions of the said definition. They can be formulated because they combine two different levels: the story level and the meta-level at which these stories find their justification as evidence of the inadequacy of Plato's definition. The story level contains narrative, plot-related elements: the state in which the featured characters—Mr. Smith, the clock watcher, the buyer of the lottery ticket—are in, and their actions. The meta-level includes events and knowledge that the characters of a story do not have access to or gain access to later: Mr. Brown in Barcelona, the fact that the clock does not work, the result of the draw.

Language (colloquial, universal) sets a trap here, one analogous to the trap that is the source of the liar paradox. To understand its mechanism, it is necessary to note the difference hidden in the approach to the concept of knowledge, just as the different ways of treating uttered sentences are concealed in the liar paradox. Knowledge is understood simultaneously as the property of a certain subject, the central figure of a story, and of another instance, whose view includes not only that figure's knowledge but also much more: knowledge of events of which the figure is unaware. To solve the liar paradox, Tarski proposes extracting formal languages that do not mix the incriminated two different levels of description from colloquial language. To resolve Gettier's problem, one must analogically acknowledge that knowledge can occur at different levels of reality and can have different instantiations. Thus, it does not belong only to the right persons who are in line with the definition and who have a justified true belief, but it also exists in a different way. For instance, it is the knowledge of the reader or the author of examples, who exist quite differently from the said persons.[5] The question of these instances has been the subject of great and intense explorations, and led to various answers at about the same time when Gettier wrote his article—although the ones we are considering here are not formulated by philosophers.

The somewhat complicated argument offered here, further complicated by the analogy to Tarski's famous definition of truth, has no ambition to speak in the

[5] An unusual interpretation, proving this fact beyond doubt and proposing a specific ontology for the area of the world in which this phenomenon appears, i.e. art in its broadest sense, is Roman Ingarden' idea of intentional beings (Ingarden 1947).

philosophers' discussion of the Gettier problem. He reaches a point that was announced earlier: certain conclusions about knowledge related to his article in which this problem was described. They are as follows. Two examples are cited in the aforementioned article, which—by way of proof by contradiction—justify the inadequacy of the definition of knowledge offered by Plato through the mouth of Theaetetus. Extrapolating the three conditions determining the occurrence of knowledge that Theaetetus deems sufficient, Gettier describes two situations in which his central figure, despite fulfilling these conditions, lacks knowledge in the subject about which they form their judgment. This means that the accepted definition of knowledge is incorrect.

And this is the point where we abandon Gettier's article. However, the analysis of his examples can go on. It shows that the knowledge which these two examples address has different instances. Speaking more precisely, in addition to the knowledge of the central figures of these examples, there is another order of knowledge, a meta-level knowledge: it is the knowledge of the central figures' knowledge. This duality is also properly realized at the level of the utterance (text), which is contained in at least two ontological orders: the reality described in the text (fiction) and the reality of the text itself (reader, author). We have proposed an analogy in this regard to the situation of truth spoken through sentences as described by Alfred Tarski. It also applies to consequences: following Tarski, one has to accept two types of language—the object language and the metalanguage—in order to create a definition of truth (and solve the liar paradox). Gettier's examples are marked by the existence of similar, analogous levels. There is the subjective knowledge and the knowledge about knowledge, respectively, which raises the question of the possible instances of this knowledge. The former has its subjective disposer. The owner of the latter is unclear, indeterminate—to say the least. If determining them is possible at all.

The Problem of Text

Reasoning based on insights into the textual background of Gettier's objections finds strong support within the domain of textual and language theory, for it is precisely text and language that provide the tacitly overlooked and yet absolute experiential basis. After all, the environment of analysis is sentences, arranged into texts (utterances, statements). This circumstance also makes a strong case for the analogy of this environment to the conditions Tarski adopted in studying the problem of truth. And it is extremely easy to find support for this because one can refer to a very well-known idea put forward by Umberto Eco in his book *Opera aperta* (The Open Work), published in 1962 (Eco 1962). He elaborated on this idea later on in e.g. *Lector in fabula* (Eco 1979). It is part of a whole system of philosophical reflection that has spread in many directions, encompassing art, science, and political and social reality. This system is known as e.g. poststructuralism, which refers to its theoretical and speculative part, and postmodernism, which comprises a certain broadly understood practice, combining artistic and social undertakings.[6] Hermeneutics provided one of the most important foundations for both of the said names, and the text became a subject of unusually intense and revealing interpretations in which sophisticated analytical apparatuses were exercised and the applications thereof were extended as the text expanded as a universal pattern or metaphor. Among most of the best-known participants of this movement, populated mainly by French philosophers such as Barthes, Foucault, Lyotard, and Derrida, the eminent philosopher and literary scholar Umberto Eco played a special role. His thoughts should be considered fundamental in the area he dealt with, which is the analysis of text understood as a reflection of the world in all its possible meanings, in which he consistently maintained both a descriptive, research approach and, in keeping with his own thoughts, a creative one, full of new ideas.

Eco's most important such discovery and—at the same time—proposal was the idea of an open work, this 'opera aperta' contained in the title of his book, which acts as a great source of inspiration and support in our case. The idea is actually obvious in its simplest formulation, but this simple discovery blossoms and grows, as is usually the case with similarly eminent thinkers, into a distant and extraordinary reflection. Eco begins this reflection as follows:

> In fact, the form of the work of art gains its aesthetic validity precisely in proportion to the number of different perspectives from which it can be viewed and understood. These give it a wealth of different resonances and echoes without impairing its original essence. [...] A work of art, therefore, is a cornpiece and closed form in its uniqueness as a balanced organic whole, while at the same time constituting an open product on account of its susceptibility to countless different interpretations which do not impinge on its unadulterable specificity. Hence, every reception of a work of art is both an interpretation and a performance of it, because in every reception the work takes on a fresh perspective for itself (Eco 1989, 3).

Eco writes about any work of art, not just literary works. Yet, the scholar remained close to the literary domain all his life, both as a theorist and as an author—the author of many well-known novels and short stories. Text is definitely where he feels most at home, and it is from the windows of this home that one can best see the distant landscapes of the other arts and sciences. What he writes about the work of art is a story also—or perhaps mainly—about literature, as his later books, such as the aforementioned *Lector in fabula*, prove. But the broader perspective, encompassing all art, is important because it supports the dynamics of the work of art's existence which he describes and which is of such great significance here. A work of art, he writes, is an object composed for its audience. This definition makes it a very broad category. A work of art happens between people in such a way that impulses flow from it towards the recipient and encounter the sensitivity and intelligence of the latter, producing themselves anew each time in this extraordinary act. It's a universal process that occurs whenever a situation fulfills these very general conditions. Of course, it is now only a step away from a very general understanding of a work of art, which is actually the everyday, which was a separate discovery of

[6] In the Polish tradition, this last area of thought has been called 'ponowoczesność', meaning 'postmodernity'.

twentieth-century art. In light of the reasoning discussed here, the most important thing is to introduce the viewer to the process of a work of art's coming to being. From now on, no sculpture, no piece of music, no text can be treated as artifacts that are independent, complete, and exhaustive in all of their qualities. Although this thought, this premonition, appears in a number of reflections, Eco expressed it with particular emphasis. This, of course, is also a deformation of some common idea of the object and its way of being.

It's a strange event that destroys the independence, the sovereignty of an object, usually considered so far as exceptional—after all, a work of art seems to be exactly so. It takes away its exclusive existence, tearing down, in fact, its ontological foundations in the process. Eco's examination of this event continues in *Lector in fabula*, which he writes already as a semiologist. The book focuses exclusively on text, with the artistic context of its origins becoming weaker—or abandoned altogether. This is understandable because the intention of the author (creator) loses its significance since their work comes to being on the basis of a universal mechanism that always works—not only when the author is an artist. Eco begins his argument with the familiar and very good known observation that "il testo é una macchina pigra che esige dal lettore un fiero lavoro cooperative per riempire spazi di non-detto o di giá-detto rimasti per così dire in bianco, allora il testo altro non e che una macchina presuppositionale" (Eco 1979, 26) ("the text is a lazy machine that demands from the reader a proud cooperative work to fill spaces of unsaid or already-said left blank, so to speak, then the text is nothing more than a presuppositional machine"). But it contains two elements that become not so obvious by being placed side by side: the unspoken and the already spoken.

This innocent connection foreshadows the violent explosion of circumstances and conditions under which the meaning of a text is produced, which draws on a variety of resources and strategies to materialize itself for the reader—whose participation proves to be extremely demanding and far from easy. Eco here advocates an approach in which the text as a physical materialization of a certain system of language, an event occurring in a certain place at a certain time, meets the active properties of that system, which, in turn, anticipates this physical materialization in various ways (it simply anticipates what it should be like in order to say what the author intends). It can do this because, after all, it learns its own uses and adapts continuously to all situations in which it has already appeared as a concrete text that has met with certain reception.[7] The game that emerges is based on a very complex strategy on the one hand, and is extremely creative on the other. The text, Eco writes,

---

[7] In order to make it happen, it uses three different repositories of meaning. The first is the situation—certain possible circumstances in which the text may appear. These circumstances are encoded in the language and do not relate to the physical realization of the text, such as an utterance that takes place at some time and under some physical conditions. They are the language's knowledge of how its structures can be applied in practice. Then, there is the context, which is the linguistic environment of individual words that form a certain semantic system with it. The third repository, limiting the context, selecting only what actually appears in the text from it, was referred to by Eco as the co-text.

constantly anticipates the behavior of the audience, drawing them into its own game. At the same time, it allows for constant modification and adaptation of the language system, absorbing the current circumstances in which it is materialized in every act of its implementation. The physical reader, anticipated as much as anything in the text, plays a very important part in this process. For it is through this reader—and their competence—that language is able to adapt itself to any circumstances in which it becomes necessary to produce a text that serves and accomplishes a purpose, and to transform it accordingly if necessary—for example, to generate new words or change the meanings of the existing ones. Language, we should add, is in no way restricted in its fertility and allows for the creation of very far-reaching innovations, limited only by our imagination. And since it easily crosses the boundaries of reality, the situation calls for a concept of possible worlds parallel to this reality to describe this capacity of the text. They are filled with unusual individuals (such as Little Red Riding Hood), become densely populated with them, but still completely consistent. In this consistency, moreover, lies the secret of the paradox of fiction ('lifelike').

This is, however, an excursion going too far, although relevant to considerations regarding knowledge. For now, let us pause at an observation concerning the way in which Eco explores the text, for that is the primary purpose of referring to him and his output. This is because the text exists in a certain continuum of its update, running smoothly through all of its possible levels. This is accompanied by an ever-present awareness of a state in which the text exists, as if in its superposition, a simultaneous existence in multiple forms for different—all possible—observers. For they are not just readers. Viewed as an emanation of language, the text engages all users of the latter, expanding its own boundaries in a virtually limitless way. By analyzing the text, we delve into an infinite grid of possible instantiations of words, sentences, and utterances that have happened and that might happen. The entire past and future experience of language reflect upon themselves in the text each time. This kind of vision, on the one hand, rearranges and reorganizes our experience of the text if we wish to penetrate it meaningfully and understand what it has to say, but also—on the other hand—allows and simplifies the everyday situations of its use, prompting automatic, learned procedures of interpretation every time we see or hear it.

The text is not only lazy, as Eco writes. It is also fragile and sensitive, since there lies a complex, volatile, and sometimes non-obvious tangle of very different, egoistic references—only partially codified in encyclopedias and thesauri—behind each of its incarnations. This fragility also manifests itself in the physical process of becoming when the text emerges as a certain physical entity. This moment has been explored in depth, for example, as the opposition of orality and literacy. Walter Ong (1982) addresses this actually dramatic opposition, which we often tend to forget, at length. He also reminds us of its ancient and deep sources, which include Plato bemoaning—through the mouth of Socrates in the *Phaedrus*—the damage that writing causes. It weakens, he argues, the memory and thus the mind, and imparts an apparent and deflated existence on thought, which then becomes immobilized and unable to defend itself. Ong views this as similar to writing to a computer. He also mention e.g. the violent eruption of reflections associated with the aforementioned opposition, which took place in France in the 1960s and 1970s, involving philosophers and literary scholars such as Algirdas Julien Greimas, Tzvetan Todorov, Roland Barthes, Phillipe Sollers, Jacques Derrida, Michel Foucault, and Jacques Lacan. This is the same environment where Eco operates.

Ong explains the aforementioned dramatic nature as follows: "Without writing, the literate mind would not and could not think as it does, not only when engaged in writing but normally even when it is composing its thoughts in oral form. More than any other single invention, writing has transformed human consciousness" (Ong 2002, 113). So, the text retains its causative power to the very end, to its last trace. Ong also draws attention to the sign as a being that opens up a separate space of analysis, arguing with Derrida on the primacy of thought and writing in awakening and producing symbols that get together to form sentences. Derrida was an advocate of the paradoxical primacy of writing in this process. Invoking this field of observation, in turn, directs us toward the record, the representation, the signature, which are always ultimately the result of formal decisions based on codified systems. Writing systems have evolved in many rich forms in various parts of the world—primarily in pictorial form. The alphabet, however, as he writes, was invented only once, around 1500 BC by a Semitic people or peoples in the Middle East. Text is not an easy being. It exists in an intricate way because it is stretched between endlessly emerging circumstances. It is impossible to be certain even of the effects of its seemingly simple physical becoming.

The present argument is already heading towards its conclusion. It started from a critical point, which is Gettier's reservation concerning the definition of knowledge, valid for more than two thousand years and supported by the authority of Plato. The discussion that erupted around it proved that it was a very important moment. Not a mistake or a superficial effect of showy thinking, but a real insight into the essence of the reasoning behind the definition in question, exposing its imperfection. However, if we turn our attention not to Gettier's reasoning itself, but to the tools he uses—stories, narratives, a different, somewhat lingering, but nevertheless immediately understandable central figure emerges—the text, which is an emanation of language. Following this path made it possible to read the diversity of levels of description, or—if you like—the multiplicity of subjects revealed as a result. Since these were knowing—but at the same time described and describing—subjects, it became necessary to interpret the levels of reality to which these subjects, their knowledge, and the texts associated with them belonged. Hence Tarski and his idea of language. This argument leads inevitably to a shift from knowledge to text, and then to an immersion into its vague and intricate world that language produces of itself every time it is used. Enter Eco and later Ong with their exemplary attempts to challenge this property when traced to such an instantiation of language as text.

In a textbook on epistemology, Robert Audi speaks explicitly of the concept upon which the said property rests. Giving examples of sources of knowledge, he lists the first and most important: perception, followed by others—memory (knowledge of the past), consciousness (knowledge of the inner life), reflection (knowledge raised to the level of abstraction), and testimony (knowledge acquired by others). He writes that "in approaching these central topics in epistemology—which may be conceived as the theory of knowledge and justification—it is appropriate to begin with perception" (Audi 2003, 1), and then, providing examples, refers to humans right away. In writing how knowledge is produced, he emphasizes the three acts during which a certain self perceives, believes, and justifies knowledge. Duncan Pritchard echoes in agreement: "Two things that just about every epistemologist agrees on are that a prerequisite for possessing knowledge is that one has a belief in the relevant proposition, and that that belief must be true" (Pritchard 2006, 5). Such an explicit reference to this person, who is the relevant subject in the case of knowledge, is actually natural for the Platonic origins of thinking about it. Over time, however, it clearly becomes just a kind of assumption. And immediately afterwards, it begins to move towards a case of knowledge in which it acquires an independent, autonomous nature in the sense that it is no longer at the exclusive disposal of man. This assumption makes it rather difficult to understand the phenomenon of knowledge in the modern world—if we understand 'modern world' as the last 100–150 years, i.e. the period since the end of the nineteenth century. This phenomenon occurs because of the intense development and reconstruction of ways of understanding knowledge. And it is unique in that it's possible to interpret it openly by noticing the text and its unusual properties.

Fortunately, there is no need to create a justification for the extraordinary role of text from scratch because the matter has already been discussed at great length. In fact, nothing in this area is simple or meaningless. Ong recalls the philosopher whose contribution to the analysis of writing is undoubtedly most profound, most significant. The name he refers to is Jacques Derrida, who shared his reflections in his book *De la grammatologie* (Of Grammatology) (Derrida 1967). In the book, he presents an unusually insightful and broad concept of the mutual arrangement of basic notions such as language and writing—along with all of their developments and extensions. His concept is based on the observation of a certain historical trend, which he refers to as writing taking precedence over language, translating into a kind of reversal of roles, according to which writing ceases to play a merely secondary, auxiliary role. Instead, it produces its own game of meanings, which as if replaces and develops the game played by language. Moreover, the story of writing is also the story of science (and, obviously, of knowledge) because, as Derrida argues, in the civilizational and philosophical project of the West, these two phenomena appeared together and in combination: "Writing is not only an auxiliary means in the service of science—and possibly its object—but first, as Husserl in particular pointed out in *The Origin of Geometry*, the condition of the possibility of ideal objects and therefore of scientific objectivity" (Derrida 1997, 27). Writing demands that we accept its capabilities and requirements, which are actually met only by mathematics. Such writing requires its own science—which is grammar.

These words, spoken half a century ago, can be interpreted in a completely novel way in the context of recent experiments in natural language processing.[8] If it really is, as Derrida writes when establishing the conditions for a study devoted to writing, that "1) the very idea of science was born in a certain epoch of writing; 2) that it was thought and formulated, as task, idea, project, in a language implying a certain kind of structurally and axiologically determined relationship between speech and writing" (Derrida 1997, 27), then writing cannot be considered as a transparent vehicle. Derrida's idea provides an argument for his essential commitment to the design of science—a certain science—that lies at the foundation of Western civilization. If we put aside the historiosophical argument, which depicts a certain development and a certain confusion of the stories of Europe, we can see evidence of the extraordinary role of writing—or rather, of its persistently recurring motif as a certain construction, a certain invention, and a certain dynamics of events. The story of writing takes place at the deepest levels, inhabited by completely free currents of creation of meaningful structures used to describe the world, capturing it in very abstract or non-obvious concepts. Encumbered with these concepts, they explain themselves only within their own systems of meaning. Very distant from common experience, yet pulled by threads that bring them to the surface to find confirmation and justification of their deep connections. The places where they become visible and significant, where they are brought to light, have the value of facts, and as such are subject to the passage of time, signaling some merely current, temporary state of the meaning carried by the text. One such place, mysterious and stunningly obvious, is the concept of book. Derrida collects texts in which its metaphorical power and capacity are used—representations of the world, or—more precisely—its order and, at the same time, a way of accessing it (Derrida 1997, 16). These are examples of writing that is no longer valid. Writing that is friendly and disciplined, appealing to nature and its divine order.

Derrida by no means equates writing with text. The latter is not a frequent object of his interest—rather a secondary one, understood as the apprehension and immobilization of writing, well captured in its graphic form. As if on a continuum of three concepts: language, writing, and text, the third of them was waiting to repeat the revolution that writing had made by superseding language and subjugating it, the revolution that text would now make against writing. From the perspective of today, this seems to be the case: text appeared to be a temporary pinnacle of writing—but a text that was not uniform, full of clutter, turmoil, excess, and exaggeration. A trivial text. And this text, paradoxically, has actually revived the book—not as an insight into the order of things that has begun, but as a local handbook or guide. Offering a closure, prompts, and especially an explanation and clarification. A text that is part of a new technology that makes use of a basic communication tool—the keyboard.

[8] As I'm writing these words, the world has just become impressed by a new algorithm called GPT 3, in which an artificial neural network is based on 175 million parameters, which is ten times more than previous language models, and makes use of text in a way that seems almost no different from the human way.

Knowledge takes shape in the light of the described order of reasoning not as a judgment or belief (conviction). And in that sense, the bullet fired by Gettier proves effective. But not as a demolitionist—because I think the cannonade he unintentionally caused shifts the attention not to the target, but to the shot. It makes us look at the deadly mechanism that sets a further series of inevitable consequences in motion—the mechanism of text. It is only from this meta-descriptive level that one can see well all the components of the act of knowledge, which in this light turns out to be more of a statement, an utterance. And as a statement, an utterance, it immediately sets in motion the complex apparatus of language and text. This way of thinking about it develops into a multitude of variants, which will be analyzed in the chapters to follow and eventually lead to my main conclusion. But alongside there is another, no less dramatic belief growing alongside these variants. A belief in the need to revise judgments about knowledge, stemming from observations of its practical manifestation.

References

Audi, Robert, ed. 1999. *The Cambridge dictionary of philosophy*. 2nd ed. Cambridge: Cambridge University Press.

———, ed. 2003. *Epistemology: A contemporary introduction to the theory of knowledge*. 2nd ed. New York: Routledge.

Bauman, Zygmunt. 1997. *Postmodernity and its discontents*. New York: NYU Press.

Borges, Rodrigo, Claudio de Almeida, and Peter D. Klein, eds. 2018. *Explaining knowledge: New essays on the Gettier problem*. 1st ed. Oxford: Oxford University Press.

Burgin, Mark. 2015. *Theory of knowledge: Structures and processes*, World scientific series in information studies: Volume 5. Vol. 5. Singapore: World Scientifc Publishing Co. Pte. Ltd.

Cornford, Francis M. 1935. *Plato's theory of knowledge, the Theaetetus and the sophist of Plato translated with a running commentary*. New York: Kegan Paul, Trench, Trubner & Co.

Derrida, Jacques. 1967. *De La Grammatologie*. Paris: Les Éditions de Minuit.

———. 1997. *Of grammatology*. Trans. Gayatri Chakravorty Spivak. Baltimore/London: The Johns Hopkins University Press.

Eco, Umberto. 1962. *Opera aperta. Forma e indeterminazione nelle poetiche contemporanea*. Milano: Gruppo Editoriale Fabbri, Bompiani, Sonzogno, Etas S.p.A.

———. 1979. *Lector in fabula. La cooperazione interpretativa nei testi narrativi*. Milano: Gruppo Editoriale Fabbri, Bompiani, Sonzogno, Etas S.p.A.

———. 1989. *The open work*. Trans. Anna Cancogni. Cambridge, MA: Harvard University Press.

Gettier, Edmund L. 1963. Is justified true belief knowledge? *Analysis* 23 (6): 121–123. https://doi.org/10.2307/3326922.

Ingarden, Roman. 1947. *Spór o Istnienie Świata*. Vol. 3 vols. Kraków: PAU.

Moser, Paul K. 2010. Gettier problem. In *A companion to epistemology*, ed. Jonathan Dancy, Ernest Sosa, and Matthias Steup, 395–397. Malden: Wiley-Blackwell.

Ong, Walter J. 1982. *Orality and literacy. The technologizing of TheWord*. 1st ed. London: Methuen & Co. Ltd.

Ong, Walther. 2002. *Orality and literacy. The technologizing of TheWord*. London/New York: Routledge, Taylor & Francis Group.

Plato. 1987. *Theaetetus*. Trans. Robin H. Waterfield. Harmondsworth/New York: Penguin Classics.

Pritchard, Duncan. 2006. *What is this thing called knowledge?* London: Routledge.

Rescher, Nicholas. 2003. *Epistemology: An introduction to the theory of knowledge*. Albany: State University of New York Press.

Russell, Bertrand. 1948. *Human knowledge – Its scope and limits*. London: George Allen and Unwin Ltd.. http://archive.org/details/in.ernet.dli.2015.461169.

Shope, Robert K. 2004. The analyzis of knowing. In *Handbook of epistemology*, ed. Ilkka Niiniluoto, Matti Sintonen, and Jan Woleński, 283–329. Dordrecht: Springer Netherlands. https://doi.org/10.1007/978-1-4020-1986-9.

Sober, Elliott. 2005. *Core questions in philosophy. A text with readings*. 4th ed. Upper Saddle River: Pearson Prentice Hall.

Tarski, Alfred. 1933. Pojęcie prawdy w językach nauk dedukcyjnych. *Prace Towarzystwa Naukowego Warszawskiego. Wydział III Nauk Matematyczno-Fizycznych* VII (34): 116.

———. 1969. Truth and proof. *Scientific American* 220 (6): 63–77.

Chapter 5
Fleck and the Way the World Exists

The twentieth century was an age of unrest. Western civilization experienced a great level of overstimulation in its course and was forced to revisit and redevelop its deepest foundations. The currents of these changes formed a powerful wave that swept across all areas of reality and became a disruption from which a new state emerged, culminating in digitality. One of the myriad articulations of this extremely painful metamorphosis is the question of man's presence in and towards the world. This metamorphosis, non-conclusive and paradoxical, led, on the one hand, to an extraordinary technological advancement which, by the end of the century, caused a new digital face of the world to emerge; on the other hand, it spawned an unsatisfied sense of distrust in one's own cognitive competence. The world seemed ultimately subservient to knowledge. But it also eluded it. It must be recalled that the paradox we have just named, according to the general mechanics of the way it works, is the result of a logic that is too weak and requires a wider view—one that will offer a perspective from which it will disappear. Thus, the solution lies in understanding knowledge, because it too has played a fundamental role in the creation of the new digital world.

The world of all the entanglements in which knowledge found itself in the twentieth century saw the appearance of the figure of a Polish physician of Jewish origin—Ludwik Fleck. He is the author of a theoretical work on scientific knowledge, published in Basel in 1935 under the title *Entstehung und Entwicklung einer wissenschaftlichen Tatsache: Einführung in die Lehre vom Denkstil und Denkkollektiv* (The Genesis and Development of a Scientific Fact: An Introduction to the Science of Thought Style and Thought Collective) (Fleck 1935). Its Polish translation was published in 1986 (Fleck 1986). Its English translation—a bit earlier, in 1979 (Fleck 1979). Only then did Fleck's work become more known and recognized, which was followed by translations into other languages. In the introduction to the English edition, Thomas Kuhn describes his quarter-century-long effort to translate this landmark publication. He recalls that his first encounter with the book took place in 1949 or early 1950, and that it was actually by coincidence: Kuhn saw the title in a footnote of a publication by another author he was reading at the time. This

R. Maciag, *Knowledge as a Tale*, https://doi.org/10.1007/978-3-031-69820-0_5

title—not the reason for quoting Fleck's book—seemed interesting to him and belonged to his area of research. This research bore fruit in the form of an important and very well-known book on the mechanisms of science, namely *The Structure of Scientific Revolutions* (Kuhn 1962).

Kuhn admits in his book that Fleck's work 'anticipated' his own ideas. He returns to the unrest accompanying this vague process in 1979, writing that he remains *almost totally uncertain* regarding the inspirations drawn from Fleck (1979, viii). This circumstance, amusing because of its illogicality, is a surprising gloss not only on Kuhn's concept, but on the idea of the development of science, which sees its own reflection in this work, is rooted earlier, and develops later on. This is because Kuhn's book is a story about science, which is viewed as the expression and the field of production of knowledge. In the book, Kuhn argues that science is not cumulative but develops through revolutions from which current, prevailing paradigms emerge to explain the world and attempt, through the efforts of most researchers, to maintain internal coherence. This coherence, however, eventually succumbs to the onslaught of experience and must collapse to make way for a next paradigm, a new combination of theories and beliefs, which will also eventually share the fate of their predecessors.

However, this process is not just about the technical side of the emergence and functioning of scientific discovery. The change is extremely profound and concerns the world in which humans find themselves, desiring to get to know this world through discoveries: "Nevertheless, paradigm changes do cause scientists to see the world of their research-engagement differently. In so far as their only recourse to that world is through what they see and do, we may want to say that after a revolution scientists are responding to a different world" (Kuhn 1962, 123). Science and scientific discoveries thus turn out, unsurprisingly, to be the beginning of a certain image of the world, and this image, less obviously, is social by nature, meaning local in temporal and spatial terms. In other words, what we know produces our way of the way in which we see the world. What we are unsure of must distort this vision of the world. The world ceases thus to be absolutely stable. Anyway, Kuhn's uncertainty concerning the influence Fleck had on him can also be seen present in the world he tells us about—in his world, in which science exists in a special way. And this uncertainty pertains to the role Fleck plays in it. Fleck is positioned where the distortion occurs. So even when Kuhn grants Fleck his due subjective capacity and prepares a place for him in the history of science, this place is vague and imperfect. This surprising result is due to the nature of Kuhn's world, in which the role of what we know is well described, while the case is worse when it comes to what we are not sure os, or in other words: to what we do not know and are aware of.

Our knowledge—according to Kuhn but not only him alone—projects on the world, not the other way around. We've lost the sense that the world exists in an objective way, meaning one that draws its power from outside us. Kuhn addresses this explicitly in his book: "We are all deeply accustomed to seeing science as the one enterprise that draws constantly nearer to some goal set by nature in advance. But need there be any such goal? Can we not account for both science's existence and its success in terms of evolution from the community's state of knowledge at any given time? Does it really help to imagine that there is some one full, objective, true account of nature and that the proper measure of scientific achievement is the extent to which it brings us closer to that ultimate goal?" (Kuhn 1962, 183). Abolishing external jurisdiction in the form of a world outside ourselves (and the category of truth, by the way) is difficult to accept. The case becomes even more complicated when we think of the infinite set of complementary knowledge that undoubtedly exists, which means everything that we do not know, and the situation that arises at the point where these sets border each other. Much suggests that they produce a shared area, a kind of product, containing the knowledge of lack of knowledge. This is where Kuhn finds himself when he writes about Fleck's influence on his ideas.

Kuhn makes a step beyond knowledge—his knowledge, of course, not knowledge in general, for he has none—and exposes himself to a draft. In fact, his claim leads to the conclusion that there is nothing but draft. The abolition of the external world as an arbiter means that nothing can be fixed and certain anymore. This operation is a repetition of the ontological procedure of physics that took place at the turn of the 19th and 20th centuries, in which the concept of ether, which was to serve as a universal reference system for all physical phenomena, was rejected. The idea of relativity, which was eventually formulated by Einstein in 1905, also based on the contributions of Lorenz and Poincaré, includes in its first postulate a thesis that leads to the statement that "the idea of motion that refers to an 'absolute state of rest' is pointless" (Thornton and Marion 2004, 547). The rejection of a universal and universally valid frame of reference is, in principle, analogous to Kuhn's idea. His idea leads to the conclusion that there are different worlds for different observers. It is possible to mutually determine their position, but they do not have a common and stable point of reference. In the case of knowledge, the role of such a point of reference was played by truth, understood, for instance, classically, as the adequacy of the world and the state of mind—the understanding of this approach is dealt with by Woleński (1993). The problem is that the moment such a point disappears, knowledge must develop other procedures of its own justification. And this could be internal coherence, for instance. At the same moment, the situation of lack of knowledge, which until now was only the lack of well-established, well-grounded knowledge, changes as well. The fundamental inability to arrive at such grounding causes different instances of knowledge to demarcate their own areas of lack of knowledge—or ignorance—for themselves too. The mutual position of these combinations can only be determined at the cost of a minimum of common cognitive ground, which Fleck discovered to exist in a broadly social context. In such a world, Kuhn's 'almost total uncertainty' sounds funny. Rather, it shows that by opening the world of undetermined knowledge, one simultaneously closes the heavy door of the logical law of excluded middle. This world is actually everything that spans 'between' the extreme possibilities of what is and what is not. It is a draft where knowledge and lack of knowledge eventually mix and blend. This profound metaphysical effect, problem, and fear make themselves comfortable at the same time in the second half of the twentieth century in reflections of various types.

The dynamics of knowledge, analogous to the dynamics of systems in the theory of relativity, representing the nature of the world in its most basic layer of existence, can—and will—be a source of further consequences, those particularly concerning the way this world is described, which is the basis of the idea of discursive space presented here at the end. And this is what the argument of this book leads to. The source of these events, however, is the doubt, unrest and anxiety that accompany knowledge. They are the children of doubt sown by an observation seemingly taken for granted, one that makes knowledge hostage to social circumstances. Marian Adolf and Nico Stehr list such great classics as Max Weber, Max Scheler, Karl Marx, Karl Mannheim, Georg Simmel, and Emil Durkheim among the most important scholars who paved the way for a similar perspective (Adolf and Stehr 2014). However, the contributions of these scholars were as direct not in every case, often acting as separate major themes of their writing. This was undoubtedly the case with Max Scheler, who formulated the claim regarding knowledge in his philosophical work in a very distinct way: "In some way or other, any knowledge, and especially all common knowledge of *identical* objects, *determines* in many ways the specification (Sosein) of society. But all knowledge is ultimately also conversely determined by the society and *its* structure" (Scheler 1980, 67). Scheler, when introducing the idea of a sociology of knowledge in 1924, gave it a profound philosophical dimension. In presenting its guiding principles, he appealed to very basic and abstract principles. They described (which can be regarded as a kind of analytical preconceptions), firstly, human's existential situation, secondly, their social existence, meaning their presence in a group, and finally the theoretical possibility of describing knowledge in the context of the first two preconceptions. Thus, he considered an individual's knowledge of being a member of a group as *a priori* and antecedent, and the knowledge of perceiving others as specific to a given community. The third principle taken from the field of the theory of cognition assumed the existence of an order by means of which the existence of knowledge in the context of man and their social constructions can be described at all. Scheler did not have the temperament of a sociologist inclined towards social reality observed experimentally. The sociology of knowledge was thus born—and Scheler is credited as the author of the conceptual cluster: Wissenssoziologie—as a highly speculative discipline.

Adolf and Stehr name Karl Mannheim as the main founder of the sociology of knowledge, who, a few years after Scheler, offers a concept thereof that results from a completely different research temperament—a concept immersed in the historical context, in the time of his own political experience, and one that consciously and openly refers to Marx. A separate chapter on the sociology of knowledge concludes his book *Ideology und Utopie* (Ideology and Utopia), published in 1929 (Mannheim 1929). Mannheim is highly practical and level-headed in his view of reality; he sees it as an object of instrumental manipulation where strings are pulled by political organizations such as parties. But the problem concerns any statement about the world: "not so much with distortions due to a deliberate effort to deceive as with the varying ways in which objects present themselves to the subject according to the differences in social settings. Thus, mental structures are inevitably differently formed in different social and historical settings" (Mannheim 1954, 238). This narrative differs greatly from Scheler's hermetic and difficult reflection. Mannheim is extremely literal and direct. For him, knowledge exists as a real phenomenon that reveals itself as a statement, an utterance, and so it has its material representation, which makes it an experiential entity, close, graspable, and easy to comprehend. His political background, along with his emotional involvement, adds to his nerve and confidence, which become easily apparent and noticeable. He is actually a negation of Scheler's abstract and speculative approach. These are two different worlds: a world of detached, distant abstraction and a world of dense matter of everyday life and existence.

Adolf and Stehr do not at all mention the father of the sociology of science, who is named, on the other hand, by Bruno Latour. Sady emphasizes this sentence (Sady 2013, 211), which Latour, one of the most important contemporary sociologists of science ever, includes in his book *Reassembling the Social. An Introduction to Actor-Network-Theory* (Latour 2005). The said father of the sociology of science is our central figure, Ludwig Fleck, who is the subject of insightful research of Thomas Schnelle, who refers to him claiming that "he sociologized the theory of knowledge and the philosophy of science in order to put the philosophically problematic categories of 'reality', 'cognition', and 'truth' in a new, unified, light from this perspective" (Schnelle 1986, 263). In this view, there is a juxtaposition of important concepts—the theory of knowledge and the philosophy of science. A juxtaposition which is not necessarily that obvious. The boundary running between the two is not sharp. Schnelle actually equates them with each other, which might seem to be a mistake if one does not consider Fleck's general type of reflection. Schnelle explicitly states that he was not only a sociologist, but a philosopher. More specifically, the author of a philosophical theory of reality (Schnelle 1986, 263). From this perspective, knowledge and science, without losing their distinctiveness, identify themselves as a single type of effort which—under a variety of analytical circumstances—may specialize, develop additional features or requirements, but remain fundamentally the same in the form of the cognition of the world. However, such a perspective is, on account of its basic nature, extremely difficult and carries a risk of simplification, triviality, and similar traps hidden behind cognitive hubris. But when one looks into Fleck's output, it is clear that he achieves this level effortlessly, even though he was not a professional or even methodical expert in philosophy (Sady 2013, 213).

The concept that has become the axis of his description of scientific cognition and is best known overall is the 'thought collective'. It appears as a consequence of a hard, unambiguous, and general thesis, which reads as follows: "cognition is … not an individual process of any theoretical 'particular consciousness'. Rather, it is the result of a social activity, since the existing stock of knowledge exceeds the range available to any one individual" (Fleck 1979, 38). Fleck explains the second part of this thesis, which contains its justification, by the necessity to relativize the cognitive act, which is not a property of the cognitive subject. The subject is entangled in a certain context. Fleck offers examples thereof: a certain state of knowledge, participation in a certain cultural environment, and, finally, the use of a certain style of thought or membership in a certain thought collective. As for the latter, he writes as follows: "If we define 'thought collective' as a community of persons mutually exchanging ideas or maintaining intellectual interaction, we will find by implication that it also provides the special 'carrier' for the historical development of any field of thought, as well as for the given stock of knowledge and level of culture. This we have designated thought style." (Fleck 1979, 39).

Such reasoning does not go unpunished in the metaphysical sense because these two areas, cognition and the world, are, after all, connected by a dialectical relationship. Thus, speaking about cognition must also mean accepting certain ontological obligations that define the object of cognition—the world. And this is where we can see Fleck's philosophical temperament in full bloom; this kind of reflection opens the book and establishes the course of its argument. It is bold and direct. For it begins with the following question: what is a fact? These are the first words of the *Preface*. Fleck immediately concludes that fact is seen as a certain, stable, sovereign, and independent object of study, i.e. a judgment regarding the world. However, he moderates its lofty character by means of great sense of practicality. He puts his hands into its very matter, which turns out to be soft, malleable, viscous, heteromorphic, and therefore dependent in hundreds of ways on the grip in which it finds itself, on the vessel that encompasses it. Drawing on his chosen example of syphilis, a venereal disease, a 'disease of pleasure', Fleck describes it as a 'mystical-ethical, empirical-therapeutic, pathogenetic, etiological disease entity', touching on its diverse and not necessarily purely medical aspects. A venereal disease, embarrassing and common at the same time, encompassing intimate love stories and deeply socially empowered, as a disease full of physiological and unpleasant details, and ultimately a cause of suffering and existential dilemmas, is chosen with a ruthless, truly medical uncompromisingness.

In this hell that this disease brings upon those who suffer from it, the suspicious, difficult and sensitive, yet burning, unmissable phenomenon of fact becomes naked and shows all its shameless poverty in broad daylight. The disease, its symptoms, its course and its response to the measures taken are by no means arranged in an orderly and simple process of analysis. Rather, it is a shaky path of a continuous search for new possibilities of interpretation and action. Chaotic and diverse. In its course, the phenomenon of fact becomes disintegrated, falling into a state of changeability, fluidity, from which stronger points of support emerge, enabling the construction of a more or less stable basis for understanding. Fleck tries to isolate them; to capture the nature of their appearance in a laboratory-like, surgical, chemical manner. And after successive distillations at the bottom of the retort, he is left with the state of the relationships that bring together and connect people centered around a single problem. It is their network, which consists of a certain state of beliefs: a style of thinking which is a result of various contexts that create these relations, one that is the source of the order that will set the problem within the framework of a judgment about the world, a theory, a hypothesis.

Fact is a difficult entity. The many and diverse paths to follow in philosophical thinking on it are something that Kevin Mulligan and Fabrice Correia appear to be aware of (Mulligan and Correia 2019). This school of thought was largely started by Wittgenstein, who recognized facts as the main building blocks of the world: "The world is the totality of facts, not of things" (Wittgenstein 1974, 5). This thinking flourished as late as in the twentieth century, when a variety of interpretive possibilities became available for this concept—which itself seems to cross the boundary between the world and man. It addresses both, and without both, it actually makes no sense, bringing together the two constantly intertwining fields of ontology and epistemology. In particular, fact appears in the context of knowledge, and of its greatest challenge—truth. But there is no reason to suppose that Fleck went so far into the philosophical interpretation of fact. Yet, he was accurate in highlighting its dual nature, as if continuing and extending Wittgenstein's idea. His story of fact is an enthralling vision of the world in all its new dynamics and complexity.

This vision is precisely designed in the introduction of Fleck's work. The subject it addresses is scientific fact, chosen in a very precise manner—the relationship of Wasserman's reaction to syphilis, and the story of how it came about and what it consisted of (Fleck 1979, xxviii). The very idea that we are dealing here not with a stable, fixed element of the world that is homogeneous, but with dynamics and structure, is revolutionary. Many pages on, both questions are answered, but the answer is not that simple: "the signal [also: aviso/advice—author's note] of resistance opposing free arbitrary thinking is called a fact" (Fleck 1979, 101). This "signal of resistance"—a sign, appearing as a certain attention-drawing and concrete point of support for the reflection trying to stop and grasp a certain state—has a 'collective-thought' nature. It does not, therefore, remain at the disposal of the one who knows, the free and spontaneous mind, but comes from outside, from the world of similar minds, remaining in tacit agreement, agreed upon by other reflections of other minds, associated with each other, abiding in relationships that are ultimately social.

What is more, and what is obvious in this situation, facts are not isolated entities, but also remain in a relationship. Since they originate from a community, they inherit it, so to speak, which immediately translates into further consequences—very serious ones. They are the source of a complete world: "Facts are never completely independent of each other. They occur either as more or less connected mixtures of separate signals, or as a system of knowledge obeying its own laws. As a result, every fact reacts upon many others. Every change and every discovery has an effect on a terrain that is virtually limitless. It is characteristic of advanced knowledge, matured into a coherent system, that each new fact harmoniously—though ever so slightly—changes all earlier facts. Here every discovery is actually a recreation of the whole world as construed by a thought collective" (Fleck 1979, 102). Facts in this story are like the shifting, mutually aligning shapes that form the world. They do not come from the world, but the opposite: they produce it. This is, of course, the world of a certain group only, but sufficiently permanent to give the impression of truth.

In his vision, Fleck clearly sees facts as a certain changing and pulsating network. He notices its dynamism and mobility, but also its internal consistency. His words are not particularly scientific. Rather, they come from an alchemical formula. This impression arises perhaps from a certain naiveté of this image, perhaps from its picturesque accuracy, or perhaps owing to its magical metaphor. Rhetoric clearly dominates logic—referring to the juxtaposition that Habermas offered in describing Derrida's dilemma. Logicality is understood there as a formal order at any level of description. Rhetoric is the story itself: it is the creation of language. By moving within this ambiguity, by exploiting it, Fleck clearly—though probably not intentionally—somehow anticipates it, and gets right to the heart of one of the most important dilemmas of knowledge, which is the degree of its fictionalization. He goes on to write: "A universally interconnected system of facts is thus formed, maintaining its balance through continuous interaction. This interwoven texture bestows solidity and tenacity upon the 'world of facts' and creates a feeling both of fixed reality and of the independent existence of the universe. The less interconnected the system of knowledge, the more magical it appears and the less stable and more miracle-prone is its reality, always in accordance with the thought style of the collective" (Fleck 1979, 102).

One of the most important features of a fact is the multitude of potential networks facts it can form because their source is a particularistic thought collective, one of many. Fleck, in gathering evidence for his reasoning in the form of historical analyses of how he understood the disease he was dealing with and the entire environment of assumptions that served him, focused primarily on the asynchronous order. He described the vivid process of their development over time, which brings to mind similar stories by Foucault, Latoure, and others. But it is also easy to extrapolate the potential coexistence of collectives and their mutual interaction. The multiplicity of collectives—and the multiplicity of systems of facts that follows from the former—lead to a multiplicity of worlds made up of them. This vision may have been inspired by the ideas of Leon Chwistek (1921), whom Fleck knew personally, as described by Schnelle and Sady. Chwistek also wrote one of many, but probably the only truly scholarly review of Fleck's book.

The relaxation, loosening, or rather complete abolition of ties with the so-called real world—although in this context, such a formulation is the name of a certain fiction—has consequences also in terms of the understanding of the concept of truth, of which Fleck writes as follows: "Such a stylized solution [to a particular problem—author's note], and there is always only one, is called truth. Truth is not 'relative' and certainly not 'subjective' in the popular sense of the word. It is always, or almost always, completely determined within a thought style. [...] Truth is not a convention, but rather (1) in historical perspective, an event in the history of thought, (2) in its contemporary context, stylized thought constraint" (Fleck 1979, 100). One can recall here William James' pragmatist definition of truth, which he sees merely as "the expedient in the way of our thinking" (James 1907, 150).

Fleck adjusts his peers' views on the fundamental questions of knowledge in one go, and the depth and principality of his approach in fact abolish the need for any further detail. With one reservation. Fleck's remarks refer to scientific research procedures, but these procedures produce a certain picture of the world, so their effects are far-reaching. Fleck sees them as a stubborn, chaotic struggle with reality, but their consequence is a separate, sovereign construction composed of facts, forming a world that is, as if parallel, and one of the possible variants, surrounded by other variants. Second, the facts that make up this world form a kind of highly interrelated structure that transforms as its individual components are modified. It is therefore an image of a complex system. Fleck could not have known such a model of the world as a structured scientific proposal, for it did not crystallize until the 1970s. A disciplined notion of a system as an organized whole, which emerged in the 1940s, did not exist in his era either. The final and most essential element of his idea is: a scientific fact is not an independent fragment of reality, but rather that random place where the attention of the researcher, driven by the built-in biases of his cognitive abilities, has stopped. These biases are, in turn, the result of local and actually trivial conditions imposed by a certain community and its habits. They model a fact in a way that fits the deep-rooted expectations of that community and the mode in which it narrates the world.

The claim made in this context regarding Fleck viewed as the father of the sociology of science, an honorable one, especially coming from such an authority as Latour, does not, however, reflect the ontological and epistemological breadth of his claims, which make up an extensive philosophical system, as Schnelle points out. Fleck depicts scientific knowledge as an almost pure social construct, explaining how it emerges without falling into the paradox that would see it as merely a certain fiction. The world described by science is not fiction, which does not mean that it is an isomorphic representation of reality or even grants direct access to it. It is a certain variation on its theme, an improvisation made according to one's background and competencies, which are ultimately determined by social circumstances. This way, Fleck not only opens up the basic paths of thinking of the twentieth-century sociology of science and knowledge, but at the same time fills them with content. From the point of view of this text, he is also the primary lawgiver of a world whose new and strange landscape unfolded before the eyes of bewildered seekers of truth. He disappointed and grieved them, causing them to abandon their claims to it. Thus, he also undermined the basic postulates and ethical beliefs related to science and—in broader terms—to knowledge, which is still valid today and becomes visible and even dangerous when one traces the contemporary mechanisms of the creation, manipulation, and instrumentalization of knowledge, as covered by e.g. Magala (2021).

Knowledge as a Social Being

The discovery of the social nature of knowledge, waiting for a long time to become common, had, however, another and no less fundamental significance. It can be said that the subversion proposed by Popper in his *Logik der Forschung: Zur Erkenntnistheorie der Modernen Naturwissenschaft* (The Logic of Scientific Discovery: On the Epistemology of Modern Science) (Popper 1935), published at about the same time as Fleck's book, the idea of falsificationism, consisting ultimately in the abolition of certainty of the image of the world painted by science, and was merely an indelible blemish on the eyepiece of the research apparatus. It was actually a painful mockery of the efforts of science, which can only be sure of itself when it is wrong, because constructive proposals achieve the status of probable hypotheses at best. Fleck discovered that (scientific) knowledge lies even further from the real world. It creates its own worlds, inspired by reality and related to it—but specific and complete. Both the latter feature and the familial ties that link scientific constructs to the real world are spaces of possibility, not clear-cut determinations. This breaks the continuity of the epistemological process hitherto unthreatened in its natural—seemingly inherent—relationship with the world. Of course, the world, since it was undermined in its knowability, became immediately distant and foggy. Knowledge, in turn, apart from its social entrapment, gained in lightness and freedom, which was sealed by the defeat of the idea of truth as the ultimate guarantee. Stripping knowledge of that privilege it had enjoyed at least since the times of Plato and the definition he formulated in *Theaetetus*, made it a kind of artifact, according to the Latin root of the term: artificial, made and equated thus with other artifacts, which ultimately abandoned any transcendent foundation. Knowledge, deprived of its uniqueness, could henceforth be treated as an ordinary utilitarian object with its own purpose and use.

This shift in the treatment of knowledge appeared in a literal way first as a certain interpretative political idea, which will be addressed soon, and then as a completely utilitarian and practical approach, treating knowledge as a certain concrete resource, and then as a pretext for behaviorist interpretation, which will also be addressed in a moment. Franssen et al. provide a very interesting clue to explain the intensity and ultimate success of this turn of events. They do so when conducting an analysis that follows a slightly different path, with other concepts incorporated into the field of sociology—namely technology and the philosophical reflection thereon. In doing so, they point to a certain confusion between the closely related and constantly diffusing fields of the abovementioned technology, science, and knowledge, which eventually gave rise to a field of research called Science and Technology Studies (STS) in the 1990s, the basic principles of which were formulated by Steve Woolgar in 1991 (Woolgar 1991). What fueled these events was, among other things, World War II and the rapid development of science in the decades that followed, which sparked governments' interest in and attempts at controlling the process. These decades were governed by state-building, conservative, and right-wing sentiment as a cure for the world collapse just experienced. These manifest themselves in attention to state policies in this regard, scientometrics, and similar issues, culminating in the launch of the journal *Social Studies of Science* in 1971 (Franssen et al. 2016).

Indeed, the 1970s prove to be a watershed year for the sociology of science, which is also the study of the sociology of professional knowledge, following a somewhat different path than common, popular knowledge. Yet, the points of their intersection prove dramatic. This is demonstrated by Jean-François Lyotard in his 1979 book *La condition postmoderne* (The Postmodern Condition) (Lyotard 1979), which has become famous more because of the term 'postmodernism' used in its title[1] than because of the fact that its sequel was titled *Rapport sur le savoir* (Report on Knowledge).[2] The notion of the sociology of knowledge becomes stable, more concrete, taking on a valid—though also strongly contested—form, devoid of specific constraints, relating e.g. only to its scientific version. A 2010 compendium on epistemology describes it as follows under an encyclopedic, comprehensive entry: "the aim of the sociology of knowledge is to locate whatever body of belief a group accepts as a true account of reality, and then try to illuminate it by reference to social variables. In the first instance the questions are:

1. What exactly is believed?
2. How is that belief distributed, e.g. who believes and who doesn't?
3. What are the sources of its credibility?
4. How is it defended against doubt and anomaly?" (Dancy et al. 2010, 744).

The author of this entry is David Bloor, an emblematic figure in the sociology of science, one of the co-founders and main exponents of the so-called *strong programme*, a school of thought named after its place of origin, the Edinburgh School, affiliated mainly with the University of Edinburgh. In 1976, Bloor published Knowledge and Social Imagery (Bloor 1976), where he teaches the basic principles of his approach. Its impact may be evidenced by the fact that it was republished in 1991 and supplemented by the author only in response to the criticism it had received (Bloor 1991). Bloor offers a rather firm position: "instead of defining it [knowledge—author's note] as true belief—or perhaps, justified true belief—knowledge for the sociologist is whatever people take to be knowledge. It consists of those beliefs which people confidently hold to and live by" (Bloor 1991, 5).[3] This definition by no means refers to the narrow area of scientific knowledge, but embraces it as if in its entirety as a single fundamental social event.

There is no doubt that the idea of relativizing knowledge in the face of social contexts, its numerous entanglements in historical or political circumstances, must trigger discussions where the stakes are, after all, the foundations of human existence in the world—now fundamentally undermined. Books such as those by Merton from 1973 (Merton 1973), by Woolgar and Latour from 1979 (Latour and Woolgar 1979), and by Karina Knorr-Cetina from 1981 (Knorr-Cetina 1981) first contest the naive notion of knowledge and then analyze the mechanisms of its production in detail, from among which its symbolic and mythic habitat emerges: the laboratory. But beyond the correct, orthodox approach to this issue—i.e. that of the sociology of knowledge, a project of quite different significance appears, reversing the previous arrangements of concepts and notions. It presupposes the need for a new social science—one that will be able to bear such a burden. This conclusion actually seems necessary if we consider the fundamental nature of the problems of

[1] Translated into Polish as 'ponowoczesność', meaning 'postmodernity'. This conceptual ambivalence is a separate matter, discussed expertly by Anna Burzyńska (2006).

[2] Chap. 5 covers the figure of Lyotard in more detail.

[3] In translating this definition overtly referring to the definition of knowledge offered by Plato in *Theaetetus*, I use the version of the translation proposed by Władysław Witwicki (Platon 2002).

the legitimacy of cognition and the existence of the world as knowledge about it. Bruno Latour, the originator of this revolution, in describing the theory called ANT—Actor-Network-Theory—which lies at the heart of this project, argues as follows: "The field of science studies is often presented as the *extension* of the same normal sociology of the social to a new object: scientific activities. [...] According to this view, laboratories and research institutes were no more than the next items in a list of topics to be tackled by using the normal ingredients of social methodology that had been used elsewhere 'with so much success'. [...] But since I am trying to define ANT, I have to show how it emerged out of sociology of science by drawing extreme conclusions not only for science but also for social theory. ANT is not the branch of social science that has succeeded in extending its methods to scientific activity and then to the rest of society, but the branch (or rather the twig) made of those who have been thoroughly shaken when trying to give a social explanation of the hard facts of science" (Latour 2005, 94).

Thus we have come full circle, although we are obviously on a different, higher turn of the spiral. For we have returned anew to the fact that Fleck considered the building blocks of worlds. In this sense, he was indeed the proper father of the sociology of science, including—and perhaps especially—that based on the new social theory that Latour wrote about. It is a task of great magnitude, leading to the aforementioned reversal of concepts, to profound consequences in thinking about the co-presence, or co-existence, of the human and the world that is the non-human. Ultimately, the primacy of human causality was abolished and things were equalized in this area. All are immersed in each other's busy, intense, and constant interactions. Such an image also reverses the path of reasoning, which Latour addresses as follows: "the social cannot be substituted for the tiniest polypeptide, the smallest rock, the most innocuous electron, the tamest baboon. Objects of science may explain the social, not the other way around" (Latour 2005, 99), adding further that "scientists made us realize that there was not the slightest chance that the type of social forces we use as a cause could have objective facts as their effects" (Latour 2005, 100).

This Copernican breakthrough, in which human beings lost the privilege of causality and facts (objects) became capable of producing the social rather than merely participating in the social life of people and being subject to its consequences, paints a picture of facts functionally opposite to that described by Fleck. Latour's idea is undoubtedly the most advanced approach to social sciences themselves, and although it turns their structure upside down and changes the view of what is at their core—the social. But this social, even though understood in a new way, still remains the fabric of the world and science at the same time. The social does not cease to be difficult to grasp; on the contrary, it becomes even more elusive, more complicated. Which means that the situation of knowledge, which has to acquire new forms in order to satisfy also to some extent the non-human community and enable facts to reveal themselves, becomes more complicated as well. Latour's dash lies in the fact that he reconstructs the world in a new shape from the very beginning—for example, by referring to the classical but reinterpreted dichotomy of culture and nature. Of course, we're still talking, in fact, about knowledge, mainly scientific knowledge, which enables this dichotomy and fills it with color. This is what *Nous n'avons jamais été modernes—essai d'anthropologie symétrique* (We Have Never Been Modern) (Latour 1991) and *Politiques de la nature. Comment faire entrer les sciences en démocratie* (Politics of Nature) (Latour 1999) are about. However, since both parts of this dichotomy must reconstruct each other's position and their own meaning, knowledge must be given a new meaning. And this is what Latour is trying to do: rearrange the figures of concepts and notions (used to describe the world) on a certain modified board that illustrates their relations and possible displacements. This way, knowledge acquires a new vocabulary that defines the very possibility and the conditions of its existence. In fact, never before have we been further from a sense of what we understand as knowledge than in this case right here.

Latour brings Fleck's revolution to a paradoxical end. In the great Frenchman's world, a fact does not act as a mere "signal of resistance" which simply stops travelers and makes them look at a landscape that would otherwise turn into a blur outside the window. It is resistance in itself, which additionally crosses the boundary of passivity and makes any further movement impossible. Fleck's and Latour's visions share an unexpected similarity of a thick, lively organism of facts, although in the latter's case, facts are more like a voracious herd than a spatula-pushed suspension. In both cases, however, knowledge—about facts—gets entangled in the dynamism of facts, and there is no single best vantage point. Nor does it offer anyone such a vantage point, for the observer is, in both cases, only a participant, speaking on their own behalf and for their own cause as if. Latour emphasizes the reflexive—because it is talking about oneself, about one's own procedures—and relativized—because there is no universal and stable observation point—nature of reflection on knowledge, which is not an easy feat. He goes on to write: "And yet, for the first time, social scientists had to study something that was *higher*, *harder*, and *stronger* than them. For the first time, the *explanandum* resisted and grinded the teeth of the *explanans*' cogs to mere stumps. Not only that, but the screams of those being studied could be heard loud and clear — and they were not coming from Bali, the ghettos, TV studios, corporate board rooms, or the US Senate, but from departments next door, from colleagues in the very same hiring and grant committees" (Latour 2005, 98). For these colleagues and entire departments have been expelled from their Olympian heights and thrown into the abyss of everyday life, along with its grim reality of a social fabric that is not nice and polite, but actually poisonous, disgusting, and teeming with the instinct for survival.[4]

This story, the subject of which was the sociology of knowledge and/or science, in which Fleck plays a significant role of the one who sowed the seeds of the revolution headed by Latour to the end of the world as we knew it, leads to a clear conclusion for our reasoning. The insignia of our power over knowledge has slipped from our hands (if we ever possessed them at all), and the world (of facts) has turned out to be a violent jungle which we control only partially, or perhaps even not at all.

[4] This is the accidentally born alternative image of the reality of science today, existing underneath a layer of official courtesy. An excellent analysis of its dynamics was offered, for instance, by Bourdieu (1992).

What we call knowledge and remains at our disposal is, was, and will be only a small and uncertain fragment. Another conclusion elaborates on this condition. That knowledge which is available must be stripped of the dogma of infallibility and turn into a usable object.

Meusburger proposes a useful distinction between two types of knowledge, which may be useful here: *factual knowledge* and *orientational knowledge* (Meusburger 2015). The former is "the sum of what has been perceived, discovered, or learned by means of a methodically well-regulated procedure bound to justification, truth, and verification" (Meusburger 2015, 24). He refers here to Abel's analysis (Abel 2008) featured in a book starting an entire series on knowledge — of which he is also the editor. Orientational knowledge is more difficult to define. It is definitely not subject to the duties of factual knowledge. On the contrary: its matter is interpretation, providing justifications and values. It feeds on myths, legends, cultural rituals, and does not necessarily appeal to the intellect. Meusberger derives these two types of knowledge from the much older categories of *logos* and *mythos*. Using such a context highlights the fundamental nature of the issue under consideration here.

Fortunately, we have an excellent analysis of the mutual relationship existing between these categories. An analysis carried out by the eminent expert on the subject, Władysław Stróżewski. In 2008, he published an article titled *Logos i Mythos* (Stróżewski 2008), conducting an insightful study that draws on multiple interpretive traditions. The researcher offers eventually a concise interpretation, which is quite an achievement in light of the old Greek — and early — sources of these concepts. I would like to cite it almost in its entirety on account of its poignant clarity and completeness, retaining the original division into paragraphs:

> The reality in which we live is not accidental: a hidden *logos* exists, which gives sense to everything that exists and to its development.
>
> This *logos* is not fully accessible for our cognition and our understanding of reality. On the other hand, it is the light through which we can understand anything. [...].
>
> We face logos as the task consisting in the postulate of discovering it. It is manifested in different ways. Mythos is the name we give to the result of our way of understanding logos.
>
> The varied ways of experiencing logos constitute the variety of mythoi, for instance: the mythoi of religion, philosophy, particular sciences, poetry.
>
> Consequently, we face mythos as a task—the task of understanding it. This is where the need for hermeneutics arises. This is why we construct theologies to elucidate the mysteries of the Holy Scriptures, or the history of philosophy, to penetrate the sense of its systems and its development, and the hermeneutics of text, particularly of poetic text.
>
> Thus, our task consists in reaching logos through mythos, returning to the source. It is one of the most important paths towards its discovery—maybe the only one. The circle is complete (Stróżewski 2008, 325).

Thanks to Stróżewski's analysis, the distinction made by Meusbereger receives an unusually deep, metaphysical—by definition—sense. But, at the same time, it brings one closer to a proper understanding of the processes we are dealing with. And it ultimately sums up knowledge-related efforts as follows: Logos remains and will remain hidden. This ancient belief marks the boundaries of any possible knowledge. In a trivialized form, stripped of metaphysical depth, subjected to ultimate secularization, and reduced to the level of social materiality, it leaves orientational knowledge as the only accessible knowledge: mythos—a story that toils, gains in layers, and explains itself in its efforts to approach the world.[5]

It is quite amusing that Stróżewski manages to offer something like a return to conviction, to concern about cognitive competence—albeit under completely different analytical conditions and using different concepts. Logos, appearing as a Heraclitean construct from the early days of philosophizing in the European circle, placed man in a similar situation as the constructivist idea of knowledge, which originated in the second half of the twentieth century. It is a situation of inaccessibility to the world as such. Meusberger, encountering and identifying the ambivalence of logos and mythos with the ambivalence of factual and orientational knowledge, however, does not mention this (he probably believes in the availability of factual knowledge). Yet, in both juxtapositions, this circumstance becomes highly and powerfully evident. What changes is its interpretation. It becomes secular, and the divine jurisdiction is replaced by the jurisdiction of social processes. This fundamental swap of interpretations between analytical and cognitive conditions of the world's existence does not resolve the fundamental problem of the question about the world—the answer to which is supposed to be knowledge. It brings it down to a level that is potentially testable—also empirically. A wicket of promise opens, leading one not directly to the mystery of the world, but to its social image.

This social image of knowledge carries a powerful charge of mere practicality, which makes it very valuable. This is the knowledge that the great surveillance projects of the early twenty-first century will ultimately compete for using sophisticated computing techniques such as artificial neural networks, resulting in a rapid race we can see today. This image will also help construct new models of how societies function. Because of this practicality, knowledge will be objectified and instrumentalized. It will turn into an artifact, a product, a thing, which will ultimately materialize in the form of knowledge as a spare part in technical engineering structures. The technology that will master this capability will be digital information technology. But then something else will also happen: freed first from its divine sanction, then from its social sanction, knowledge will turn out to be independent of human presence. It will be the greatest revolution in the understanding of knowledge since Plato. Which is happening as I'm writing these words.

References

Abel, Günter. 2008. Forms of knowledge: Problems, projects, perspectives. In *Clashes of knowledge. Orthodoxies and heterodoxies in science and religion*, Knowledge and space, ed. Peter Meusburger, Michael Welker, and Edgar Wunder, 11–33. Dordrecht: Springer.

[5] Equating scientific knowledge with literary art is thus permitted. Science can even be considered a kind of writing; this belief emerged in the second half of the twentieth century among philosophers of language, which will be covered more extensively in Chap. 4.

Adolf, Marian, and Nico Stehr. 2014. *Knowledge*. Abingdon: Routledge.
Bloor, David. 1976. *Knowledge and social imagery*. London: Routledge & Kegan Paul.
———. 1991. *Knowledge and social imagery*. 2nd ed. Chicago: University of Chicago Press.
Bourdieu, Pierre. 1992. *Les Règles de l'art: genèse et structure du champ littéraire*. Paris: Seuil.
Burzyńska, Anna. 2006. *Anty-Teoria Literatury*. Kraków: TAiWPN Universitas.
Chwistek, Leon. 1921. *Wielość rzeczywistości. Wielość rzeczywistości*. Kraków: Min. W. R. i O. P. [Wyznań Religijnych i Oświecenia Publicznego].
Dancy, Jonathan, Ernest Sosa, and Matthias Steup, eds. 2010. *A companion to epistemology*. Malden: Wiley-Blackwell.
Fleck, Ludwik. 1935. *Entstehung und Entwicklung einer wissenschaftlichen Tatsache: Einführung in die Lehre vom Denkstil und Denkkollektiv*. Basel: Benno Schwabe.
———. 1979. *Genesis and development of a scientific fact*, ed. Thaddeus J. Trenn, Robert K. Merton, and Thomas S. Kuhn. Trans. Frederick Bradley. Chicago: University of Chicago Press.
———. 1986. *Powstanie i rozwój faktu naukowego: Wprowadzenie do nauki o stylu myślowym i kolektywie myślowym*. Trans. Maria Tuszkiewicz. *Powstanie i rozwój faktu naukowego: wprowadzenie do nauki o stylu myślowym i kolektywie myślowym*. Człowiek i Świat Człowieka. Lublin: Wydawnictwo Lubelskie.
Franssen, Maarten, Pieter E. Vermaas, Peter Kroes, and Anthonie W.M. Meijers, eds. 2016. *Philosophy of technology after the empirical turn*, Philosophy of engineering and technology. Vol. 23. Cham: Springer International Publishing. https://doi.org/10.1007/978-3-319-33717-3.
James, William. 1907. Pragmatism's conception of truth. *The Journal of Philosophy, Psychology and Scientific Methods* 4 (6): 141–155. https://doi.org/10.2307/2012189.
Knorr-Cetina, Karin. 1981. *The manufacture of knowledge: An essay on the constructivist and contextual nature of science*. Elsevier Science Limited.
Kuhn, Thomas S. 1962. *The structure of scientific revolutions*. Chicago: University of Chicago Press.
Latour, Bruno. 1991. *Nous n'avons Jamais Été Modernes – Essai d'anthropologie Symétrique*. Paris: La Découverte.
———. 1999. *Politiques de La Nature. Comment Faire Entrer Les Sciences En Démocratie*. Paris: La Découverte.
———. 2005. *Reassembling the social: An introduction to actor-network-theory*. Oxford: Oxford University Press.
Latour, Bruno, and Steve Woolgar. 1979. *Laboratory life: The construction of scientific facts*. Princeton: Princeton University Press.
Lyotard, Jean-François. 1979. *La condition postmoderne. Rapport sur le savoir*. Paris: Éd. de minuit.
Magala, Sławomir. 2021. *The third enlightenment (or globalizing meritocracies)*. Cambridge Scholars Publisher.
Mannheim, Karl. 1929. *Ideologie und Utopie*. Vol. 3. Bonn: F. Cohen.
———. 1954. *Ideology and utopia: An introduction to the sociology of knowledge*. Trans. Louis Wirth and Edward Shils. London: Routledge & Kegan Paul.
Merton, Robert K. 1973. *The sociology of science: Theoretical and empirical investigations*. University of Chicago Press.
Meusburger, Peter. 2015. Relations between knowledge and power: An overview of research questions and concepts. In *Geographies of knowledge and power*, ed. Peter Meusburger, Derek Gregory, and Laura Suarsana, 19–74. Dordrecht: Springer. https://link.springer.com/chapter/10.1007/978-94-017-9960-7_2.
Mulligan, Kevin, and Fabrice Correia. 2019. Facts. In *The Stanford encyclopedia of philosophy*, ed. Edward N. Zalta, Winter 2019. Metaphysics research lab, Stanford University. https://plato.stanford.edu/archives/win2019/entries/facts/.
Platon. 2002. *Parmenides, Teajtet. Przełożył oraz wstępami i objaśnieniami opatrzył Władysław Witwicki*. Kęty: Antyk.
Popper, Karl. 1935. *Logik der Forschung: Zur Erkenntnistheorie der Modernen Naturwissenschaft*. Vienna: Springer Vienna.

Sady, Wojciech. 2013. *Spór o Racjonalność Naukową. Od Poincarego Do Laudana*. Toruń: Wydawnictwo Naukowe Uniwersytetu Mikołaja Kopernika.

Scheler, Max. 1980. *Problems of a sociology of knowledge*. 1st. ed. London/Boston: Routledge Kegan & Paul.

Schnelle, Thomas. 1986. Ludwik Fleck and the influence of the philosophy of Lwów. In *Cognition and fact. Materials on Ludwik Fleck*, Boston studies in the philosophy of science, ed. Robert S. Cohen and Thomas Schnelle, vol. 87, 231–266. Dordrecht: D. Reidel Publishing Company. https://doi.org/10.1007/978-94-009-4498-5.

Stróżewski, Władysław. 2008. Logos i mythos. *Przegląd Filozoficzno-Literacki* 2-3 (20): 311–326.

Thornton, Stephen T., and Jerry B. Marion. 2004. *Classical dynamics of particles and systems*. 5th ed. London: Thomson Brooks/Cole.

Wittgenstein, Ludwig. 1974. *Tractatus Logico-Philosophicus*. 2nd ed. London/New York: Routledge.

Woleński. 1993. *Metamatematyka a epistemologia*. Warszawa: Wydawnictwo Naukowe PWN.

Woolgar, Steve. 1991. The turn to technology in social studies of science. *Science, Technology, & Human Values* 16 (1): 20–50. https://doi.org/10.1177/016224399101600102.

Chapter 6
Masuda and the Abandonment of Man

The central figure to the beginning of this chapter is Yoneji Masuda, author of the rather remarkable book *The Information* Society *as Post-Industrial Society*, which is a description of the ambitious plan created for Japan by an independent non-profit organization—a plan that intended to transform Japanese society into a form called the 'information society'. Masuda was the main architect of this plan, which was presented to the Japanese government in the early 1970s. The book was published almost a decade later and is a conceptually mature description of the said form. This is a very special book, which is already implied by its cover where two of the most important charts illustrating Masuda's concept are pasted on a background featuring a mountain resembling Mount Fuji. Both diagrams depict the social transformations taking place: the first is the evolution of the Triple Concept of Human Society, the second is the Computer-Communication Revolution and Its Societal Impact. Such a juxtaposition, especially with the Japanese sacred mountain in the background, immediately introduces the very basic and general nature of Masuda's proposal, which includes a futuristic vision of a new type of society. Astutely and somewhat self-critically, he called it 'Computopia', predicting it to become reality in the twenty-first century. He described it as follows: "And if the highest stage of industrial society is the high mass consumption society, then the highest stage of information society will be the global society futurization society, a vision that greatly expands and develops Smith's vision of a universal opulent society; this is what I mean by 'Computopia'. This global futurization society will be a society in which everyone pursues the possibilities of one's own future, actualizing one's own self-futurization needs by acting in a goal-oriented way. It will be global, in which multi-centered voluntary communities of citizens participating voluntarily in shared goals and ideas flourish simultaneously throughout the world" (Masuda 1981, 147).

Masuda has equipped this plan, remarkable and impressive in its scope, with a meticulous analysis of the processes of its formation and functioning. Two of them are featured on the cover, aptly representing the author's organized and disciplined way of thinking. They are important also from the point of view of the present text

R. Maciag, *Knowledge as a Tale*, https://doi.org/10.1007/978-3-031-69820-0_6

(especially the second of them) because they contain the notion of knowledge. In addition to automation and the innovative transformation of the social system, knowledge is one of the three main components of the revolution brought about by computers and their impact on the domain of communication. Knowledge is also depicted there as a part of the changes taking place and as the result of a certain situation. Moreover, it can be described by the functions it serves. The latter include solving complex problems, improving the process of goal achievement, and increasing employment and educational opportunities. Thanks to this model, knowledge reveals its purely utilitarian and practical side, and is very much related to the human being who makes use of these qualities.

This model, this chart, which also appears inside the book in the chapter on the social transformation brought about by computers, completes the image of knowledge, whose definition appears a little earlier: "«Knowledge» *is no more than cognitive information that has been generalized and abstracted from an understanding of the cause-and-effect relations of a particular phenomenon occurring in the external environment.* […] We can now give a more complete definition of 'cognitive information'»: *it is an informed situational relation between a subject and an object that makes possible the action selection by which the subject itself can achieve some sort of use value*" (Masuda 1981, 55). The main element of its understanding, rooted in the subject, the human being, is still there—but its realization changes significantly. It is literal, pragmatic, and addresses the immediate circumstances of an individual's social life in a Marxian way: by emphasizing the role of work. It seems, however, that despite Masuda's clearly present historiosophical reflection, which refers mainly to the economic aspect (as exemplified by references to Smith, for instance), his thinking is determined more by his practical orientation than by his political temperament. The two cannot be completely isolated from each other, of course. Also, this relationship clearly accompanies the idea of information society, which is strongly emphasized and extensively described by Webster (2006).

Information Society and Post-Industrial Society

There is a fascinating story to the concept of information society regarding its birth, or—to be more specific—a story of a certain rivalry regarding the primacy of its creation. The parties competing for the credit are two countries: Japan and the United States. Duff, who wrote a well-known book on this type of society (Duff 2000), begins the story of its history exactly so. He describes the process of the formation of this concept in Japan in detail, where, traversing various linguistic 'beings', it reaches its final—stable—form. Citing Youichi Ito's research, Duff states, beyond any doubt, that said society was born in Japan in 1964, when it was referred to in the composite form 'Joho Sangyo Shakai' (industrial information society) (Duff 2000, 4). A great part in its emergence was played by another researcher and scholar, Tadao Umesao, whose text published the previous year started a discussion which sparked widespread interest in the concept of *Joho Shakai* (or *Johoka Shakai*) in Japan.

Now, we come across a complication. One that leads us towards our interests and towards Masuda's concept. A year before Umesao's article, a book by Fritz Machlup that brought a reflection on the same subject was published in the United States. We can leave the heated debate themed with the question of whether the Japanese knew the book aside. It does, however, appear to turn the spotlight to the problem we are dealing with here, as illustrated by its very title: *The Production and Distribution of Knowledge in the United States*) (Machlup 1962). This brings the United States to the forefront of the said race for primacy, but the title also points to the beginnings of the use of the concept of knowledge in a way that moves completely away from its primarily philosophical—previous—use. It needs to be added here that the victor in this race was Peter Drucker, who took up the subject of knowledge, as he himself writes (Drucker 1993a, 26), in his 1961 text *The Technological Revolution; Notes on the Relationship of Technology, Science and Culture* (Drucker 1961), later reprinted in two places (Drucker 1970, 1993b). However, I have not been able to confirm this information elsewhere. Nor is it mentioned by Duff, which makes it rather necessary to recognize the primacy of Machlup's well-known and impactful concept—one much more extensive than Drucker's.

It is not particularly complex either. The problem of knowledge is just leaving the realm of sophisticated and intricate speculation and entering economic reality. At first, this transition takes place at a great level of generality, as a consideration of a certain model of social organization and—naturally for this level—political reflection. Duff highlights the importance of Machlup's work to the idea of information society and cites evidence of the widespread influence of his thought. This is fostered by the fact that Machlup treats the concepts of information and knowledge as very close, almost interchangeable, also offering an extensive explanation of his approach. Ultimately, he even recognizes that "all information in the ordinary sense is knowledge, though not all knowledge may be called information" (Machlup 1962, 15). There is a certain lightness and 'laid-backness' to this approach, which would rather unlikely be accepted per today's standards.

Machlup deals with the notion of knowledge in a rather broad sense, but this is not for the most part a reference to existing reflection—rather his own thoughts and impressions. They are structured around the anxiety of having to define a concept which is the axis, the core of the book. However, in the end, Machlup does not provide such a definition, but contents himself with a kind of classification of knowledge, which turns out to be very diverse and changeable. This abundance is eventually distributed into five main types of knowledge: practical knowledge, intellectual knowledge, pastime knowledge, spiritual knowledge, and unwanted knowledge. The first of them has even further variants: professional knowledge, business knowledge, workman's knowledge, political knowledge, and household knowledge. Spiritual knowledge includes religious knowledge and knowledge leading to salvation. From a certain perspective, this set seems rather random, and individual elements of it can be questioned. However, it emerges here as some evidence of an understanding of the problem of knowledge. It surely assumes a functional nature, which is closely related to a certain practical life—also when a religious aspect appears in it. This is actually a list of situations in which one has to demonstrate one's knowledge and which bring this knowledge in the form of experience (even if this experience is purely spiritual) at the same time. A similar approach also governs Machlup's reflection on the relation of knowledge to truth in the remaining part of his argument.

Machlup's deliberation eventually leads to the point where he concedes the presence of yet another type of knowledge—a type that crowns and closes this deliberation: "One more classification must be made, one which is essentially based on economic principles. Knowledge can be classified either as a final product or as a necessary requirement—cost element—in the production of other goods and services" (Machlup 1962, 29). And thus the most prominent, most important articulation of the concept of knowledge emerges. And it is so because it is the foundation of the book and emphasizes the most significant process of Machlup's contemporary economics: the production of knowledge. Knowledge is treated like any product: its producers and consumers can be identified, its paths of circulation in the social and economic structures can be identified, and—finally—it is subject to the power of indicators, including price. Knowledge becomes object-like, materializes itself, incorporating itself directly into economic processes and the network of economic sectors just like other commodities. And this is a highly groundbreaking perspective on the matter. Up to then, no one had had the courage to treat knowledge in a similarly concrete and direct manner. Machlup is very proficient in his argument. He offers it in a well-structured, reasoned, and extensively analyzed manner. In doing so, he creates a new economic reality, but after all, in a moment, it turns out that the consequences of his actions are much further-reaching. He also changes the idea of product, which doesn't have to be tangible in as literal a way as before. Thus, he opens up a world that is our everyday reality: the world of digital virtuality. At the same time, he makes a violent reconstruction of the phenomenon of knowledge, which irretrievably loses its deep nature, a nature contained in man's epistemic predisposition, turning into a thing, similar to other things—all the things that surround this man and organize and sort out their everyday life.

Then, just over a decade later, probably the most important book that takes way we view knowledge further is published, continuing and elaborating on Machlup's idea. This book is Daniel Bell's powerful, nearly 500-page work *The Coming of Post-industrial Society. A Venture in Social Forecasting* (Bell 1973). In keeping with its title, its main idea is quite remarkable for Bell presents his self-designed and self-justified concept of *post-industrial society* as "a social forecast about a change in the social framework of Western society" (Bell 1973, 9). This change is expected to occur within 30 to 50 years, so it is essentially a twenty-first century forecast. From the point of view of orthodox science, which, after all, aims primarily to describe and learn about the world, this goal is quite controversial. Bell understands this view very well, devoting a separate part of the book to explaining the principles of his research. In particular, it emphasizes the peculiarity of the concept of 'social framework',[1] meaning the basic and all-encompassing order of the construction of the said society. To fathom the way in which this framework exists, he goes very deep, for he invokes the nature of knowledge itself, which appears as the application of pre-prepared cognitive tools—such as language—to describe reality. These frameworks are therefore only conceptual schemas enabling us to talk about the world, offering a certain logical order through which the order of facts is somewhat revealed. Logical order exists as prior categories or even conventions that let us name the elements of that order, seen as a network of relations that make up the world. Thus, by explicitly referring to Kant, his concept of the category of intellect and the mode of cognition, Bell moves towards his own definition of knowledge. It is a kind of matching of observations with pre-existing descriptive categories prepared in advance: "it is a function of the categories we use to establish relationships such as perception" (Bell 1973, 9). Conceptual schemes are thus only incomplete by definition, constructed for a certain purpose, and their truthfulness cannot be judged for they are not judgments regarding reality; they can only be useful or not.

By offering such a notion of knowledge, Bell avoids the naive argument against his idea of social forecasting. His forecasting does not predict the future because it does not refer directly to reality, but is positioned at the level of intentional and abstract cognitive constructs. As if a level higher in the order of cognition and knowledge, where the very nature of that cognition becomes important—and not just the world as its object. Forecasting involves a skillful, experience-driven selection of categories and orders to determine the main directions of interpretation and axes of reasoning. A thought following in these directions, along these rather than other axes, and confronting them with observed phenomena, can establish certain basic insights that interpret these phenomena. This is how knowledge is born.

Here Bell applies a procedure based on the Kantian observation of the empowerment—but also entanglement and confinement—of man in *a priori* forms of cognition that both enable and determine their perception of the world. However, he interprets this empowerment somewhat 'technically', abandoning the philosophical and speculative perspectives. He is, after all, a researcher of social phenomena, which are strongly integrated into the world of facts. He therefore adopts an intermediate solution, which appears to be, in fact, an extremely modern one. The descriptive construction he proposes is only a certain proposal, justified and based on previous research, but its nature is that of a specially prepared instrument which is used only when it's ready. And so, Bell does not make the world and experience his direct associates, even though he does explore them. Instead, he gives this role to the effects of his tools. He does not rely on the appearance of a rock to judge its hardness, but watches how it behaves when treated with a chisel, which fragments are carved out, how deep the resulting fissure is, and how much force is required to use the tool. Much depends on the chisel, not just the rock. The former has the

[1] The Polish language does not have a good equivalent of the term *framework*, and translating it—out of necessity caused by the lack of a better equivalent—through the word 'framework' is far from the original, which also includes the internal, basic structure, and thus refers to a certain structural whole of the phenomenon it describes.

advantage of being produced by its user, the explorer, the investigator, the researcher, which makes it, by definition, closer and familiar to them. Although the metaphor seems to end here because, in fact, Bell's 'chisel'—meaning the idea of a 'conceptual schema'—is entirely different in terms of its matter from a 'rock'—meaning experiential social reality. It's an encounter of logic and facts. A rock and a chisel are not separated by such an ontological difference.

The idea of treating knowledge as a set of initial assumptions, despite its various philosophical filiations, shines with its own pure and remarkable brilliance. This is because it lets us give an original answer to the question of what this conceptual scheme, derived by Bell from three fundamental aspects of society as such, really is. It consists of a social structure, which includes economic, technological, and employment matters; politics, which regulates the distribution of power and the processes of power ownership; and culture, which is the domain of symbols and meanings. The concept of a post-industrial society is set upon the foundation of the first aspect—the social structure—and points to the process of change that its components undergo. It is in this space that the three main problems society as a whole, in all its aspects, faces arise. First, these problems concern the system of social roles in which individuals can and must fulfill. In this case, Bell uses the example of the bureaucratization of science, which leads to its specialization and segmentation, which is somewhat similar to the situation of industrial specialization during the industrial revolution of the 19th and early 20th centuries. Second, there is a tension between the social structure and the political sphere, which has to cope with the growing social self-knowledge and rely on knowledge in a technical and utilitarian way, which leads to a growing role of its workers: scientists, engineers, technocrats. This tension causes problems of fundamental importance within the concept of post-industrial society. And third, new patterns of living are emerging, and they are also based on knowledge—both practical and theoretical.

Being a logical construction, the concept of post-industrial society is internally clearly organized on a higher interpretive level as well. Bell precisely arranges successive levels of analysis, moving from a methodological, self-reflective foundation through the idea of society to a layer of the core axes of his reasoning—which he refers to as dimensions. There are five such dimensions that capture the organization of post-industrial society. The very notion of dimensions is rather metaphorical, as Bell uses it interchangeably with the term 'components'. They consist of: (1) economic sector, shifting from the production of goods to a service-based economy, (2) occupational distribution, dominated by technical and professional classes, (3) axial principle, which is a centrally located, theoretical knowledge that is the source of innovation and which provides policy solutions, (4) expectations of the past founded on the control of technology and technological capabilities, and (5) decision-making mode, based on the new "intellectual technology" being produced.

Bell applies the latter term to a paradigmatically different understanding of reality, in which all organizations, institutions, companies are viewed as complex systems, calling for adequate models, abandoning simple causal analyses, and appealing to advanced probabilistic techniques, to set theory, to game theory, and especially to the idea of sovereign and free systems of arbitrary assumptions introduced by Peano and Hilbert as axiomatic systems. Bell is clearly fascinated by this newfound view, where the phenomena of the world emerge as the effects of multifaceted interactions, eluding simple, unambiguous explanations, which means entirely new analytical challenges in an area as practical as the economy.

Among Bell's abovementioned ideas, the notion of knowledge is constantly repeated in various configurations. Bell emphasizes its importance again and again, making it a central and fundamental component/dimension. Explaining its role, he writes as follows: "A post-industrial society is organized around knowledge for the purpose of social control and the directing of innovation and change; and this in turn gives rise to new social relationships and new structures which have to be managed politically. [...] What is distinctive about the post-industrial society is the change in the character of knowledge itself. What has become decisive for the organization of decisions and the direction of change is the centrality of theoretical knowledge—the primacy of theory over empiricism and the codification of knowledge into abstract systems of symbols that, as in any axiomatic system, can be used to illuminate many different and varied areas of experience" (Bell 1973, 20). This kind of description becomes then the subject of a much more dramatic supplementary argument, which we'll return to shortly. But even now we can see the most important concepts and ideas, which give testimony to Bell's unusual approach. First, he upholds the idea of knowledge, whose defining property is 'theoreticality', which he clearly understands as a certain specific superiority to the field of experience. Such knowledge, not necessarily universally comprehensible because of being written in 'abstract symbol systems', is applicable to experiential practice, but it is something distinctly different from the act of experiencing. This is his epistemological contribution, shaped by experiences with the domain of mathematics. Second, this knowledge proves to be an effective tool to learn about the world through its own constructs, although it cannot be expected to provide a universal view of the world. This seems to be where its potential effectiveness shows: its local, detailed, casuistic nature brings it, quite paradoxically, closer to reality because it is always mindful of its problematic (local) cognitive status. For we are not dealing with a situation in which knowledge is a cumulative reflection of a world of unequivocal certainty which though seems to be the ultimate goal—even though it may be difficult to achieve. These are only particularistic and local insights gained using specialized but questionable ('theoretical') tools. Such knowledge lacks an initial, comfortable confidence, trust. It must constantly justify its own accuracy—of which it is never certain.

Here, Bell draws far-reaching conclusions from reflections on the status of scientific knowledge, with these reflections being represented in the most monumental, emblematic form in the analysis offered by Popper (1935). Yet, he is not interested in epistemology, but in experiential practice—or rather its social mode, which he views extremely instrumentally. This approach becomes evident later in the book, in a special chapter on knowledge. It needs to be stipulated that the definition he proposes is made for the purposes of this very chapter, which is also a materialization, a realization of the aforementioned locality of knowledge, but the way it is constructed and the conceptual apparatus used therein are in perfect harmony with the factual and pragmatic approach discussed just now. He writes: "Knowledge, then, is that which is objectively known, an intellectual property, attached to a name or group of names and certified by copyright or some other form of social recognition (e.g. publication). This knowledge is paid for—in the time spent in writing and research; in the monetary compensation by the communication and educational media. It is subject to a judgment by the market, by administrative or political decisions of superiors, or by peers as to the worth of the result, and as to its claim on social resources, where such claims are made. Thus knowledge is part of the social overhead investment of society; it is a coherent statement, presented in a book, article (or even a computer program), written down or recorded at some point for transmission, and subject to some rough count" (Bell 1973, 176).

This highly objectifying view of knowledge is actually the only one conceivable in a situation where knowledge does not claim, usurp the right to represent the world, but comes to being as a result of purposeful and useful activity. Such knowledge is not a representation of truth, but an image of practical reality. It is only that which is 'known' and not that which 'exists'. And so, it is natural for it to have modes of verification other than the existing truth-related and transcendent ones. Bell refers to official, bureaucratic, collective verification, which implies incorporating knowledge into the space of social phenomena and relations in a very pragmatic, real, even political or commercial sense. The outcomes of this approach later on became the subject of a serious debate and a poignant analysis by great thinkers. It has been addressed by e.g. Bourdieu (1992) and Derrida (2001). Not only has this problem not been resolved, but it continues to define the social situation of science. Bell further keeps on arguing that, in light of the accepted mode of verification of knowledge, what is analyzed first and foremost is the usefulness of knowledge (rather than its truthfulness). And its usefulness allows it to be valued within a framework of utility, effectiveness, and the like, which can ultimately be translated into money.

Luckily (for scientists), Bell adds that on account of the role of knowledge, it is needed, essential, and its incorporation into the utilitarian circulation of goods makes it gain in significance: "What is true of technology and economics is true, albeit differentially, of all modes of knowledge: the advances in a field become increasingly dependent on the primacy of theoretical work, which codifies what is known and points the way to empirical confirmation. In effect, theoretical knowledge increasingly becomes the strategic resource, the axial principle, of a society. And the university, research organizations, and intellectual institutions, where theoretical knowledge is codified and enriched, become the axial structures of the emergent society" (Bell 1973, 26). But the term Bell uses to describe knowledge—a 'strategic resource'—is unforgiving. Knowledge loses its innocence and eventually accepts the role of not only a desirable good but also a weapon in the no-holds-barred struggle for survival and profit. Bell addresses it directly: "The post-industrial society, it is clear, is a knowledge society in a double sense: first, the sources of innovation are increasingly derivative from research and development (and more directly, there is a new relation between science and technology because of the centrality of theoretical knowledge); second, the weight of the society—measured by a larger proportion of Gross Domestic Product and a larger share of employment—is increasingly in the knowledge field" (Bell 1973, 212).

Knowledge, which has hitherto appeared to offer the promise of its understanding, becomes ultimately secular. The act that is just happening with and through Bell's book ends its long detachment from the jealously guarded world of secrets that do not belong to man, so eloquently portrayed as a symbol in the image of the tree of news of the good and the bad. At the same time, it brings a new form of knowledge—one whose basic attribute continues to be its ownership, but this time, this ownership is exclusively human. These last words evoke a sense of sadness because the triumphant story of the Promethean dimension of this ownership is accompanied by an opposite area opening up—an area where knowledge is a record of greed, deception, and the pursuit of power. The first two are already obvious in Bell's writing. The third is passed over in silence. It does emerge as a real threat in parallel in texts by e.g. Michel Foucault or Jean-Françoise Lyotard, although it has been well known in the European cultural circle at least since the time of Plato.

The secularization of knowledge goes hand in hand, quite necessarily, with the reconstruction of the idea of man. One could say that in the surprising order and interconnectedness of discourses, knowledge, confined to a purely human legacy, needs a properly emancipated human being, and this adjustment, this adaptation has to be done only through rethinking the concept and notion in question anew. And this effort ends badly because it becomes clear that the idea of man is merely some artificial invention, devoid of any transcendent foundation. This is how posthumanism views man, where posthumanism itself is built on the progressively deeper acts of critique of humanism, put forward by Nietzsche, Heidegger, or Foucault (1966; Heidegger 1949; Nietzsche 1878). The latter ends the book with a poignant metaphor for the fragility of a human invention, made in Europe, in a fairly recent past, which is as impermanent as a footprint on the sand. Man, freed from transcendent obligations, appears naked in the world of purely social relations, which tend to be also political because of the gradual increase of their complex and multi-level nature. In these circumstances, knowledge emerges as the protagonist of a story about another field of purely social practice, subject to an extreme principle of pragmatism, just like economics. And perhaps even more—because this principle of pragmatism is power.

It was Plato who was one of the first to consider knowledge in the context of power in the political sense: as a condition, a reason for organization of the state, usually reduced to the superiority of philosophers over non-philosophers. Although this problem proved to be extremely challenging,[2] it clearly established the relationship between the two phenomena. This relationship, which has become a well-known platitude, was expressed in the words of Francis Bacon and later echoed by his secretary Thomas Hobbes. The former, who made the problem of science and knowledge the main area of his writing, is also the author of the idea of reforming them and reconstructing the ways of learning and getting to know reality, which he

[2] Addressed by Smith (2000).

covers in his *Novum Organum* (1620). It was inspired mainly by the advances in science the author witnessed in the process of writing, meaning in particular the practical inventions such as the compass, printing, and gunpowder, which turned cognition toward nature, demanding a balance between empirical and speculative approaches (Leśniak 1961), paving the way for modern science. In 1597, Bacon published a collection of essays, *Meditationes Sacrae*. It is there that the prototype of the maxim later featured by Hobbes in its Latin version in *Leviathan* (1658), widely repeated thereafter, appears: *scientia potentia est* (Rodrígez García 2001).

Despite the banality of this platitude, it points to one of the most important centers that accumulate reflection and constantly inspire researchers (see: Gilder 2013). The aforementioned Rodrígez García connects Bacon directly with possibly the most important thinker to devote a large part of his work to the issue in question—Michel Foucault. The interdependence of the issue of knowledge and power is considered one of the major themes of Foucault's writings, although the issue of knowledge in the humanities appeared in his works long before he formulated his questions of power (Rouse 2005). It should also be added that he was undoubtedly motivated to take up this subject matter by the political events of '68, which have deeply penetrated the fabric of reflection and affected a significant number of research contexts. The amplitude of the latter went from, for instance, political or moral practice through social and cultural reflection to sophisticated philosophical constructs and, in horizontal order, spanned a great many academic disciplines (Foucault 1980). In this sense, it was a revolution that could not be ignored. This revolution is accompanied by the blooming of a concept of knowledge that will impose itself with dramatic force on all later authors addressing the issue.

References

Bell, Daniel. 1973. *The coming of post-industrial society: A venture in social forecasting*. 1st ed. New York: Basic Books.

Bourdieu, Pierre. 1992. *Les Règles de l'art: genèse et structure du champ littéraire*. Paris: Seuil.

Derrida, Jacques. 2001. *L'université sans condition*. Paris: Galilée.

Drucker, Peter. 1961. The technological revolution: Notes on the relationship of technology, science, and culture. *Technology and Culture* 2 (4): 342–351. https://doi.org/10.2307/3100889.

———. 1970. *Technology, management & society: Essays*. New York: Harper & Row.

———. 1993a. *Post-capitalist society*. New York: HarperBusiness.

———. 1993b. *The ecological vision: Reflections on the American condition*. New Brunswick: Transaction Publishers.

Duff, Alistair S. 2000. *Information society studies*. Abingdon: Routledge.

Foucault, Michel. 1966. *Les mots et les choses. Une archéologie des sciences humaines*. Paris: Gallimard.

———. 1980. *Power/knowledge. Selected interviews and other writings 1972–1977*, ed. Collin Gordon. New York: Pantheon Books.

Gilder, George. 2013. *Knowledge and power: The information theory of capitalism and how it is revolutionizing our world*. Washington, DC: Regnery Publishing, Inc.

Heidegger, Martin. 1949. *Über Den Humanismus*. Frankfurt am Main: Klostermann.

Leśniak, Kazimierz. 1961. *Franciszek Bacon. Myśli i Ludzie*. Warszawa: Wiedza Powszechna.

Machlup, Fritz. 1962. *The production and distribution of knowledge in the United States*. Princeton: Princeton University Press.

Masuda, Yoneji. 1981. *The information society as post-industrial society*. Washington, DC: World Future Society.

Nietzsche, Friedrich. 1878. *Menschliches, Allzumenschliches: Ein Buch Für Freie Geister*. Chemnitz: Veralg von Ernst Schmeitzner.

Popper, Karl. 1935. *Logik der Forschung: Zur Erkenntnistheorie der Modernen Naturwissenschaft*. Vienna: Springer Vienna.

Rodrígez García, José María. 2001. Scientia Potestas Est—Knowledge is power: Francis bacon to Michel Foucault. *Neohelicon* 28 (1): 109–121. https://doi.org/10.1023/A:1011901104984.

Rouse, Joseph. 2005. Power/knowledge. In *The Cambridge companion to Foucault*, ed. Gary Gutting, 2nd ed., 95–122. Cambridge/New York: Cambridge University Press.

Smith, Nicholas D. 2000. Plato on knowledge as a power. *Journal of the History of Philosophy* 38 (2): 145–168. https://doi.org/10.1353/hph.2005.0082.

Webster, Frank. 2006. *Theories of the information society*. 3rd ed. Abingdon: Routledge.

Chapter 7
Foucault and the Riddle of Discourse

Here begins the story of Michel Foucault, for whom knowledge is a fundamental problem and who is the most important figure in our argument. It is his analysis that makes it possible to organize knowledge and—leaving behind the structuralist objections—to introduce a much more advanced order to the world, which, as it turns out, was already present in the original. However, it is a story so vast that it deserves a separate chapter. So let this chapter be an inner whole of the larger part that opened (and will end) with the story of Yoneji Masuda.

Knowledge, however, is the most essential problem for Foucault. In fact, he already explores it in a systematic manner in his early book *Folie et déraison. Histoire de la folie à l'âge Classique* (*M*adness and Civilization: A History of Insanity in Age of Reason) (Foucault 1961). This very rich and vivid story of ways to perceive, qualify, and deal with madness includes a great number of documents and testimonies, producing a sufficient collection to move to a more general level: knowledge, its mechanisms, and nature: "the consciousness madness, in European culture at least, has never formed an obvious and monolithic fact, undergoing metamorphoses as a homogeneous ensemble. For the Western consciousness, madness has always welled up simultaneously at multiple points, forming a constellation that slowly shifts from one form to another, its face perhaps hiding an enigmatic truth. Meaning here is always fractured (Foucault 2006, 163)." This observation opens a generalization that will become the cornerstone of Foucault's thinking about knowledge. Dispersion, decentralization, fragmentation, divergence—these are synonyms that can (nowadays) capture a peculiar characteristic of the way knowledge exists in Foucault's understanding. This idea is very preliminary for now, but already consciously fixed in the general context of knowledge, although for the time being, it appears as a series of questions: "But what form of knowledge, after all, is sufficiently singular, esoteric or regional to be given only at a single point, in a unique formulation? What learning could be so well—or so badly—understood to be known only in a single time, in a uniform manner, in a single mode of apprehension? What figure of science, however coherent or tight it might be, does not allow

R. Maciag, *Knowledge as a Tale*, https://doi.org/10.1007/978-3-031-69820-0_7

more or less obscure forms of practical, moral or mythological consciousness to gravitate around it? If it were not experienced in a random order, and recognized only in profile, all truth would soon fall dormant" (Foucault 2006, 159).

The history of madness is necessarily also the history of normalcy as its opposite, and therefore an in-depth and full—or even dramatic—examination of humanity. The latter's most important feature, the ultimate proof of its supremacy, is the traditionally accepted reason, but it is the latter that actually falls prey to insanity. Insanity, in turn, as if in an equally basic and final impulse to oppose and repair the order, becomes the subject of strenuous interpretive and research efforts, aimed to describe and explain it in a rational way. For Foucault, madness becomes a sign of the contradiction between the unintelligent and the rational; it is a hiding and wandering non-reason, faced against rationality, all in-depth diagnoses, classifications, and etiologies, which attempt to make sense of the unintelligent and restore reason as the basis of life.

It's easy to get lost in this conundrum, which is also at the same time a heroic yet deeply pessimistic story for it shows the weakness of reason, its fragility, and the efforts to defend it often become highly desperate. However, the concept of sense and meaning pushes the matter out into yet another field, which will henceforth become the core axis of Foucault's line of thought. The circling and tracking of senselessness in a relentless manner while it tends to dissipate and hide cannot just be understood as a relentless establishment of knowledge. And knowledge gained in this way does not follow a straight and predictable path, but gets lost in the boundlessness of the combed area, subject to fantasies and entangled in the intrigues encountered.

The history of madness, based on Foucault's experience as a psychologist, the author of treatises that are the grounds of his credentials in psychology and psychopathology, stops at this preliminary view of knowledge. To him, reaching the truth seems a busy activity, always incomplete, unfinished, and complex, one that does never yield an image that is singular, stagnant, immobile, immersed in lethargy, but an image that comes in a variety of versions and fragmented at the same time. Perhaps it is this approach, directly related to the human being, the patient, the suffering being, that defines Foucault's point of view and prompts him to the area in which he seeks to root knowledge. The 1960s is a period of an intense, growing interest in language, which begins much earlier and is finally 'decreed' by Rorty as a *language turn* in 1967 in his collection of philosophical texts by various authors.

Their purpose, as Rorty writes, is to provide texts for reflection on the latest philosophical revolution—a philosophy of language, which chooses language as the primary interpretive category (Rorty 1992, 3). The origins of the renewed interest in language in the twentieth century, which continues till this day, go back to the end of the previous century—the works of Frege, Wittgenstein, Russell, which share a common feature in the form of the pursuit of cold, logical elegance and order, gradually loosening up with the encountered failures of such an approach. In the 1950s, in Wittgenstein's argument, language ceases to be a logical picture of the world, and takes the shape of a metaphorical, labyrinthine tangle of the streets of an old city (Wittgenstein 1958, 8). A transcript of lectures given in 1955 by John Longshaw

Austin at Harvard is published in his book *How to Do Things with Words* (Austin 1962) in 1962. In this remarkable and very original work, the author makes an extensive analysis of speech acts, relying on colloquial language taken from everyday experience to prove its extraordinary ability to create reality.

The interest in language as an object and tool of philosophy offers a large area for reflection, where various competing views clash. This area is also 'inhabited' by Willard van Orman Quine, author of a work published in 1960 and titled *Word and Object* (Quine 1960). He belongs to the ranks of logicians, and although the aforementioned book begins with the words "language is a social art" (Quine 1960, ix), its argument is cool and in formal in a number of places. It fits in well with the characteristics offered by Merquior, who, in describing the origins of Foucault's style of philosophy, places it in the Gallic tradition of philosophical practice, so different from the Anglo-Saxon and the German-speaking schools, which are both academic and analytical (Merquior 1985). Foucault and his linguistic circle remain closer to literature, their texts are rich in impressive and varied vocabulary and are full of splendor that rises above discipline and strictness. There is surely a lot of truth in this statement, although it's hard to accuse Foucault's book of lack of discipline, in which the issue of knowledge—and in broader terms—the activity of knowing and describing the world, is addressed as if forcefully, highly extensively and insightfully, making it perhaps the philosopher's most famous and influential book. Its title paraphrases the title of Quin's book—although I found no direct confirmation of this claim, which is implied by their similarity in both time and thematic scope. *Les mots et les choses. Une archéologie des sciences humaines* (The Order of Things. An Archaeology of the Human Sciences) (Foucault 1966) was published in 1966.

Deleuze describes this work as follows: "As Foucault explains, *The Order of Things* deals neither with things nor with words. It is not concerned with object or subject, nor does it examine phrases or propositions, grammatical analysis, logic or semantics. Statements are not in any sense portrayed as a synthesis of The Order of Things, or as composite phrases or propositions. On the contrary, they precede the phrases or propositions which implicitly presuppose them and lead to the formation of words and objects" (Deleuze 1988, 12). The critical aspect here is the notion of 'statement', which is the axis of the book's subject matter.[1] The concept is far from easy. On the contrary, it is very problematic and basic at the same time. In the English language version, published two years after the French original appeared, it is translated from the French 'l'énoncé'. The phenomenon described by this term is something prior to words, sentences, or specific utterances. Deleuze explains it as a certain profound capability to utter certain meanings that materialize as the mentioned words or sentences. A capability contained in the linguistic corpus. Its nature is difficult because it does not mean a specific linguistic form, known from common experience, but a structure that organizes these words and sentences, forcing them to express it. However, since these words and sentences are the only form of

[1] The same concept concluded Chap. 1.

manifestation of this statement, it remains dependent on these manifestations, bound to them in a dialectical and—as Deleuze claims—paradoxical relationship. There's no clear relationship between statements and words and sentences. They're not arranged in a strict hierarchical order. Statements are materialized in numerous variants of word and sentence combinations, producing unique and numerous versions each time. The universe of the former is related with that of the latter. They mutually propel each other, revealing their individual distinctiveness in the pure act of speaking, when words and sentences come to being before our eyes for a moment—only to melt into a sense of statements the next moment.

The way in which these statements—the 'énoncés'—exist is very mysterious and poses a serious ontological problem, which I have decided to grapple with, led by Foucault's guidance. The results of my efforts are presented later on in the book. Deleuze doesn't make the task any easier. He points to a certain feature that turns out to be an indelible effect of their verbal, textual articulation. These statements are, out of necessity, *multitudes,* or, more precisely, *multiplicities* (orig. *multiplicité*), as Deleuze cites Riemann, a mathematician and the author of the concept in question (Deleuze 1988, 13). Immediately afterwards, he goes on to recall two more philosophers: Husserl and Bergson, who interpreted the concept—each in their own way. Yet, Deleuze claims that their ideas—unlike the said mathematical idea—failed to survive. Compared with the above, Foucault's book stands out as "the most decisive step yet taken in the theory-practice of multiplicities" (Deleuze 1988, 14). And it does so not directly, by defining the concept, but by consistently referring to such peculiar form of manifestation of the world. This way, multiplicity inevitably enters our habitat as yet another mental construct essential to the understanding of Foucault—as well as his concept of knowledge, even though he himself did not directly refer to Riemann. For multiplicity is not a concept from the dictionary of philosophers, but mathematicians. It is a naming of the order of reality based on numerosity. It is impossible to understand language as Foucault viewed it without this abstract exercise, which reveals a truth well-known today: language and the world are two completely different things, existing separately, sovereign, and complete in their being. But they are sets. Language creates, builds, and arranges its own space, that's clear. What is less clear is drawing conclusions from such a formulation when we take away its metaphorical aspect, which only serves to simplify its nature, in fact. It is analogous to the world—spatial, space-related. But what do we do when this observation is still too simple and calls for more insight?

Today, multiplicity is a formally defined mathematical entity, proposed by Bernhard Riemann in 1854 in his lecture *Über die Hypothesen, welche der Geometrie zu Grunde liegen* (On the Hypotheses Which Lie at the Bases of Geometry) (Riemann 2016). But the concept of multiplicity was presented by Riemann in a much less disciplined way than we would expect from a mathematician, which also paved the way for other kinds of interpretation. Riemann's lecture in which he introduces this concept is discussed in more detail by Torretti (1978, 67). The important context here was the emergence of a new way of thinking about geometry, exemplified by its non-Euclidean idea devised by Nikolai Lobachevsky. An idea that had been commonly known since the 1840s. Geometry gradually broke away from its real foundation and took on an uninterpreted and abstract form. The culmination of this process was Hilbert's concept of 1899, presented in his book *Grundlagen der Geometrie* (Foundations of Geometry) (Hilbert 1899). Geometry became a speculative field, detaching itself entirely from material reality, the definitional 'land measurement'.

Torretti emphasizes the groundbreaking role of Riemann's lecture, which, in order to remove problems related to the foundations of geometry, "must clarify the general idea for which space is only a special case." Unlike present-day mathematicians, Riemann considered space to have a less extensive meaning. He viewed it as a place where physical bodies are located, subject to physical displacement (Torretti 1978, 83). But this is no longer enough for him. He writes that "the general concept of multiply extended magnitudes [...] remained entirely unworked" (Riemann 2016, 31 [1854]). He considers many different spaces, and in order to do so, he must proceed to a more general, higher level of reasoning, where space is an instance of a certain *broader genus* (Torretti 1978, 84). This kind of thinking eventually leads him to propose an entity of an overarching, superior nature, which he refers to as *Mannigfaltigkeit—manifold*. The term describes a phenomenon that would today be called a set (Torretti 1978, 85). From the perspective of the time in which it is created, it is a completely new creation, and makes us think about constructing ways to describe certain seemingly obvious properties of the world in a new manner. There is clearly a tension in this reasoning, which will become a big geometric problem—which will be described in more detail in Chap. 7. It's the tension between the merely imagined but strict geometric (and topological) concepts and the real world—a necessary exercise of abstract imagination, which then returns to the world in question to describe it.

The modern approach of mathematics, seen in the efforts to define the topological manifold, which can be considered here as the main interpretive variant of manifold, is much narrower. Its formal definitions are offered, for example, by Kirby and Siebenmann or Tu (Kirby and Siebenmann 1977, 7; Tu 2011, 48). A descriptive definition of a more formal version of a manifold is provided by Lee in his textbook: "The underlying idea is that manifolds are like curves and surfaces, except, perhaps, that they might be of higher dimension. Every manifold comes with a specific nonnegative integer called its dimension, which is, roughly speaking, the number of independent numbers (or "parameters") needed to specify a point" (Lee 2011, 1). This geometric—but not at all everyday—image does not seem so difficult as long as, of course, the perspective of many, very many dimensions is not daunting.

Of the two 'civil'—i.e. non-mathematical—interpretations of manifold mentioned by Deleuze, the one offered by Bergson is of lesser interest to us in this case. The idea of multiplicity appears in his 1889 doctoral dissertation, addressed as a suitable tool to understand the peculiar nature of mental states. They follow their own path of time as simultaneous and consecutive beings, and their own path of space as different and connected beings, which calls for a new understanding of time. Multiplicity makes it possible to describe their dialectical nature of simultaneous single and multiple existence (Bergson 1889).

Husserl approached the issue of manifold in a much more in-depth and extensive manner. Milkov, describing the idea of manifold at length, to which Husserl refers in abundance, cites the latter's definition: "Manifolds are thought, infinitely open, sets of objectivities *[Gegenständlichkeit*—author's note] that are defined by axiom-forms, i.e., by forms of laws that are thought as valid for objects, or for their otherwise undetermined and formally thought combinations and relationships" (Husserl 2019, 219). This definition comes from a series of lectures the philosopher gave in late 1917 and early 1918 (Husserl 1996). Milkov's standpoint on this definition is clear: "This means that the objects in a pure manifold are absolutely undetermined" (Milkov 2005, 124).

Husserl's line of thought originates in the very generally understood idea of mathematics, which, as Milkov points out, refers to the old idea of Leibnitz's and Descartes' *mathesis universalis*. Husserl elaborates on the idea of calculus, going beyond numbers: "One can calculate with concepts and with propositions in the same way as with lines or forces or surfaces, etc. Naturally, it makes no sense to measure propositions using numerical values. The calculation is, therefore, not calculation with quantities and numbers" (Husserl 2008, 78). Mathematics grows to the size of a principal area, encompassing much more than just calculations. Husserl's reasoning is based on an approach which Milkov considers as "rather, (…) a «mathematics» (logic) of an indefinitely general realm of thinking" (Milkov 2005, 125). This is the type of approach which Husserl adopts when describing the conditions for the existence of manifolds: "one can *construct* manifolds (therefore, domains conceived in indeterminate universality that are the bases of a possible mathematics) by *arbitrary definition*" (Husserl 2008, 83).

This kind of idea is very general in nature and is devoid of any determination, such as that enforced by the material world. At the same time, it offers unlimited conditions for organizing this world—if we ever wish to do so—by making use of the symbolic representation of this world in a construct which can take various forms, including formalized ones (mathematized). This kind of assumption underlies the coherent theory of manifolds proposed by Husserl, which preserves order despite its freedom and arbitrariness: "Let there be a domain in which the objects are subject to certain forms of relation and connection, for which axioms of such and such a form are valid, then for a domain formally constituted in this way, a mathematics of such and such a form would be valid, there would then result propositions of such and such a form, proofs, theories of such and such a form" (Husserl 2008, 87). The fundamentality of this concept, its purely abstract and—in fact—logical nature, makes some authors, such as Smith, also view it as a kind of theory of everything (Milkov 2005, 125).

The latter, reconstructing Husserl's idea of manifolds, goes even further, generalizing the concept of manifold: "a manifold is the ontological form of the (highly complex) state of affairs represented by a given theory (which is a set of propositions, or their conjunction, a complex proposition)." (Smith 2002, 110). Milkov considers such an approach as purely ontological (and therefore not necessarily mathematical or geometrical), and recalls Wittgenstein's idea in this context. In such an ontology, "every state of affairs has its own manifold" (Milkov 2005, 129).

The notion of '*the state of affairs*' is a peculiar idea regarding the mode of existence, which, as Smith notices, is shared by Husserl and Wittgenstein. This circumstance becomes important in light of the fact that the latter also used the concept of manifold, although, as Milkov argues, he understood it in a very different way than Husserl, and he positioned it very differently in the general continuum of entities. Milkov summarizes his approach as follows: "Wittgenstein saw the manifold—the common element between matter and mind (language)—as an *indefinable*" (Milkov 2005, 129). The latter term referred to a special class of entities: the indefinables, which are "intuitively known simples" (Milkov 2005, 134).

The introduction of the idea of manifold when dealing with Foucault's writing aims first to show an interesting property of his reasoning, existing alongside the sensitivity to a broad social context, which is frequently emphasized in his case and which is, in fact, a sensitivity to the experiential condition. It is abstract synthetic thinking, which naturally requires certain intrinsic precision, like all inference procedures that exist outside the empirical world of reference, because it is only in this precision and its discipline that it finds justification. Within this thinking, Deleuze also finds a structure that can be interpreted as a manifold that fulfills these conditions. This is possible because there is a certain area of generalization, abstraction, and interpretation of phenomena. An area that is earlier and more basic than concrete and purposeful applications of the conceptual construction of word. It is this very area that Husserl attempts to delineate, elaborating on the concept of manifold. At the same time, it makes it possible to organize observable phenomena, which are always only examples, existing in great numbers, multiplied in the social world and producing whole galaxies of events and relations. According to Deleuze, Foucault reconciles these dialectically stretched states: the great number of occurrences and the singularity of accidents, which Foucault describes in abundance, referring to sources and accounts, and thus remaining in constant, intense contact with the world of experiences. In the order of our reasoning, this is an introduction to the constructs and ideas that will appear in the last chapters and which are apparently not foreign to the French philosopher.

For Foucault, the two great worlds of experience, where Deleuze finds a reflection of manifold and its sense as described above, are, on the one hand, language in its purely practical form, de Saussure's *parole*, and, on the other hand, knowledge, understood as all that can be contained in language, but which is obviously not language. Both of these worlds are open and unstable sets. They eventually coexist as interdependent on each other, but this relationship is paradoxically balanced by the extraordinary flexibility of the dependencies that tie them together and which are unstable and ambiguous. Foucault discovered that the resulting structure is historically changeable, although it can stabilize to some extent. Temporary cognitive constructs—*episteme* (originally in French: *l'épistémè*)—epistemological fields (i.e. fields of knowledge), vast in size and volume, come then into being. They appear to be historical and changeable, though on a grand, civilizational scale. These fields are discontinuous, repeatedly connected and highly active, lively matter, made up of smaller units that as if mediate between the levels of knowledge and language: discourses. To understand the dynamics of these connections, these linkages, it is necessary to refer to a construct that is general, or speaking more precisely, that closes the march of generalization—the manifold. Now that Deleuze has given us this ordering tool, we shall make use of the power of synthesis to enter the world of discursive phenomena that Foucault unfolds before us, writing as follows: "Once these immediate forms of continuity are suspended, an entire field is set free. A vast field, but one that can be defined nonetheless: this field is made up of the totality of all effective statements (whether spoken or written), in their dispersion as events and in the occurrence that is proper to them. Before approaching, with any degree of certainty, a science, or novels, or political speeches, or the oeuvre of an author, or even a single book, the material with which one is dealing is, in its raw, neutral state, a population of events in the space of discourse in general. One is led therefore to the project of a pure description of discursive events as the horizon for the search for the unities that form within it. [...] The field of discursive events, on the other hand, is a grouping that is always finite and limited at any moment to the linguistic sequences that have been formulated; they may be innumerable, they may, in sheer size, exceed the capacities of recording, memory, or reading: nevertheless they form a finite grouping. The question posed by language analysis of some discursive fact or other is always: according to what rules has a particular statement been made, and consequently according to what rules could other similar statements be made? The description of the events of discourse poses a quite different question: how is it that one particular statement appeared rather than another?" (Foucault 1972, 26–27).

Discourse According to Foucault

The extensive quotation just cited comes from Foucault's *L'archéologie du savoir* (The Archeology of Knowledge), another of his books, after *The Order of Things*, in which he explains his concept of discursive knowledge in a slightly different, more disciplined, detailed, and precise manner (Foucault 1969). From now on, the concept of discourse will prevail in the minds of those who wish to access knowledge through the language with which it is spoken. This is such a vast field that even a brief description of it requires a separate chapter. A discursive event can only be studied, as Foucault argues, by asking about the reason for it to happen. This is, after all, another variant of the question of the meaning of such an event, based on the assumption that statement *(l'énoncé)* always has and serves some purpose. And this is the minimum condition for its meaningfulness and purposefulness: "an 'enunciative datum' must exist that will always be determined and not infinite" (Foucault 1972, 111). Therefore, discourses are justifications for sets of relevant statements. According to Deleuze's line of thought, we should say that discourses materialize within certain spaces that exist together and in combination in certain manifolds. They exist in a dynamic manner, although it's possible to notice certain more stable states through which we gain insight into the meaning, or—more precisely—into the knowledge that is contained in these—listed successively in abbreviated form—discourses, statements, sentences, and words.

Foucault summarizes his concept of knowledge as follows: "[a] group of elements [form and rigor, objects, statement types, notions, strategies—author's note], formed in a regular manner by a discursive practice, and which are indispensable to the constitution of a science, although they are not necessarily destined to give rise to one, can be called knowledge. Knowledge is that of which one can speak in a discursive practice, and which is specified by that fact: the domain constituted by the different objects that will or will not acquire a scientific status [...]; knowledge is also the space in which the subject may take up a position and speak of the objects with which he deals in his discourse [...]; knowledge is also the field of coordination and subordination of statements in which concepts appear, and are defined, applied and transformed [...]; lastly, knowledge is defined by the possibilities of use and appropriation offered by discourse [...]. There are bodies of knowledge that are independent of the sciences (which are neither their historical prototypes, nor their practical by-products), but there is no knowledge without a particular discursive practice; and any discursive practice may be defined by the knowledge that it forms." (Foucault 1972, 182). This definition is attributive and ontological. It does not refer to the content of knowledge, so it does not have to go the trouble of establishing criteria to sort out the content, which is a challenge knowledge analysts come across quite frequently (Burgin 2015). According to Foucault, knowledge—or more precisely, the condition of its existence—is based ultimately on the existence of a discursive practice. The two condition each other and each one comes down to and is the essence of the other.

It is then clear that the key to understanding knowledge becomes discourse, the description of which, as understood by Foucault, has already appeared. Its remarkable utility has turned it into a field applied and interpreted in many different ways by other authors, to which we shall return. The story of Foucault began with a retrospective observation made by Rouse, according to which the problem of knowledge appeared earlier in the philosopher's work, followed by the question of power only later on. It also began with the author's remark that the situation may have been influenced by dramatic historical circumstances, which is to mean the events of 1968, deeply and greatly affecting the social, cultural, and all other strata. The above is evidenced in Foucault's next statement, which takes the form of an inaugural lecture delivered at Collège de France on 2 December 1970, published soon after as a separate work (Foucault 1971). Foucault makes the theme of the lecture clear from the start: "in every society the production of discourse is at once controlled, selected, organized and redistributed by a certain number of procedures, whose role is to ward off its powers and dangers, to gain mastery over its chance events, to evade its ponderous, formidable materiality" (Foucault 1981, 52). Foucault names three such procedures: (1) social exclusion, which can take different forms, but always threatens to challenge the subjective right to participate in the community and enforces certain behaviors through this threat, (2) internal control and censorship, which impose a constant reconciliation of one's own perspective and views with an oppressive and disturbing exterior, or (3) simple restriction of access, imposing a division between those entitled or not entitled to participate inside a certain exclusive discourse. The first one is most strongly concerned with the ways of exercising power, but the other two have to do with those ways as well, which eventually makes all three proceed to the level of reflection on the nature of society. And since knowledge is inherently connected to a discursive event, both remaining in a relationship that allows both to exist, knowledge is brought into the same field, becoming at the same time aware of the obvious justification for its connection to power.

In this rather dramatic situation, which turns out to be painfully real in light of the events of 1968, the need for a good explanation of the mechanisms of knowledge and power naturally grows stronger. The interpretive tools adopted call for as precise an explanation as possible, and the main, most primary of these tools is discourse. Foucault's lecture offers the most accurate description of this phenomenon to date. The philosopher gets perhaps closest to its hard-to-grasp, fleeting nature, which is subject to constant regulatory measures that are the usual result of the exercise of power. Or rather, of the application of discourse as power, which is becoming more and more apparent. This application, this dominion, so to speak, is materialized through language, whose relationship with the world becomes ever closer, but this is only an apparent victory on the road to truth (meaning the knowledge of the world). With regard to discourse, Western culture is split into two—simultaneous and opposite—currents, so to speak. On the one hand, it lets discourse work laboriously at the boundaries of thought and language in order to learn about the world, and on the other hand, it limits its space, its field in order to succeed in enclosing it in the rigid structures of meaning, which are necessary to rule over people.

This ambivalence, this alternation is the subject of an ongoing power game that has a history, but never ends—because it is sustained all the time by the dynamics of social processes. This leads to an unclear and dynamic situation of knowledge—and ruling, which Foucault describes as follows: "For it would seem at first glance that by rediscovering everywhere the movement of a logos which elevates particularities to the status of concepts and allows immediate consciousness to unfurl in the end the whole rationality of the world, one puts discourse itself at the centre of one's speculation But this logos, in fact, is only a discourse that has already been held, or rather it is things themselves, and events, which imperceptibly turn themselves into discourse as they unfold the secret of their own essence. Thus discourse is little more than the gleaming of a truth in the process of being born to its own gaze" (Foucault 1981, 65). These extremely difficult words, offered in poetic form, can emphasize—on the one hand—the essential value, the relevance of discourse, and—on the other—its use and consumption. And hence a certain rebellion and opposition in the name of discourse as a space where the most important work of reason as a field of play of thought and language can take place.

Foucault focuses in the remainder of the argument on the analysis and description of this work, formulating the method he will use for this purpose. This method is necessarily a simultaneous determination of the conditions under which we will view discourse—a creation of an entirely unclear nature. This way, it becomes visible, and the proposed tools enable us to see it—in certain characteristic features—by outlining its shape. This method is based on four principles, rules (orig. *Principes*, *règles*): reversal (*principe de renversement*), discontinuity (*principe de discontinuité*), specificity (*principe de spécificité*), and exteriority (*règle de l'extériorité*). These principles open the door to a remarkable opportunity to grasp the nature of discourse—and will prove extremely useful later on. The first one opens the field to the next one and concerns the mechanics of how stories and knowledge are produced at the same time. Its dominant quality is the painful perversity that afflicts us whenever we try to answer the most fundamental questions, with Foucault posing three of them: concerning the author, the discipline, and the truth, i.e. the subject that speaks, the mode of our speaking, and its ultimate fortune. When it comes to these very issues, we are in a particularly strong need of certainty, which is definitely not provided by discourse because it disintegrates into chaotic polyphony, condemning us to a state of eternal uncertainty. But this hermeneutic remark goes alongside a morphological statement that announces a certain fundamental and potential property of discourse, which is its nature of a dispersed and unevenly compacted set. This is, of course, some abstract approximation, a way to get some idea of the object in question, which, after all, has no ultimate, definitive ontological explanation. It comes closer and gradually takes shape precisely through similar approximations, a dynamic known well from e.g. Wittgenstein's writings, especially when he describes the nature of language in his *Investigations.*[2]

The principle of reversal is extended to include the principle of discontinuity, which treats us to an image of a free universe of discourse flows, i.e. the existence of some indefinite and extensive set of discourses, which is mobile and changeable in addition. The difference is that Foucault imbues the principle of reversal mainly with the unpredictability of the order of discourses, its changes in density, their unexpected nature, which is not subject to any external order—i.e. speculative or logical—either. This is the first observation of any new participant in this reality. A participant who has just earned the right to view its interior and realizes that they play the role of a student who can foresee little, and especially cannot expect that any of the important dilemmas they may be facing must have a solution. The second principle produces a technical view of it: "Discourses should be treated as discontinuous practices that intersect, sometimes juxtapose, but can also be ignored and excluded" (Foucault 1981, 67). But it also offers a good insight into its mechanics—a set of somewhat separate and mobile beings, which, as one might guess, come in plenty and are very different from each other. This property immediately evokes a very familiar image of a complex phenomenon, one well-described in the domain of physics. We will return to it shortly.

It is accompanied by the third principle: specificity. It is not entirely clear. It seems to be split and inconsistent, combining a technical description with a sudden self-reflection. The technical description is ruthlessly accurate and emphasizes the complete independence of this lively nebula of knowledge that defies

[2] For example, when he writes about the nature of inquiry at the very beginning (Wittgenstein 1958, vii).

comprehension. This is because discourse like world is not "the accomplice of our knowledge" (Foucault 1981, 67) and does not succumb to elaborate systems of meanings. You can't just read it, decipher it, because there is no pre-existing language that would make it possible. Discourse is, after all, above any language. The self-reflection that follows offers a clue to the just emerging question of what, in view of this, discourse is. According to this clue, it operates directly in the world. But it is not a fortuitous action. Rather, it is desperate and doomed to failure, because it is an act of "violence which we do to things" (Foucault 1981, 67). Discourse is amorphous in the sense that it is not organized in some reasonable, purposeful way, it is not locked within a structure of meanings. It is the opposite: it produces meanings. If it is knowledge, it remains in a state of potency which is about to become reality and materialize itself in the form of words. Discourse mimics the nature of quantum physics beings, which exist as simultaneous, all, and infinite of their possibilities, without losing the peculiar nature of a particle of one kind or another. Discourse is similarly paradoxical. And so it must be since knowledge is also produced as its effect—not just its cause. In light of the above it is therefore impossible to know what came before.

The fourth principle, exteriority, is perhaps the most rigorous, stern, aimed directly at the thoughtless daredevil who ventures into the universe of discourse. The previous rule, specificity, has already begun using the language of commands after all, but this time these commands are not concerned with the rather vague space of meanings, and thus regulate a movement that is only metaphorical. This time it is a command referring strictly and decidedly to geometric relations, even though it is abstract geometry. "Ne pas aller du discours vers son noyau intérieur et caché" (Foucault 1971, 55)—"we must not go from discourse towards its interior, hidden nucleus" (Foucault 1981, 67)—warns Foucault, taking on the role of a cautious guide who protects explorers against getting lost and shows the right way, acting as a rather strict, demanding teacher. The instructions are clear and protect those who follow them against the threats awaiting on the wrong path: "but, on the basis of discourse itself, its appearance and its regularity, go towards its external conditions of possibility, towards what gives rise to the aleatory series of these events, and fixes its limits (Foucault 1981, 67). Discourse is deceptive and misleading, able to clearly deceive the beholder with some mirage of a momentary cluster of words and meanings that will obscure others. Thus, it is not where the discourse needs to go, where it becomes concrete, specific, and where it takes on a recognizable shape, but precisely in the opposite direction: where it just emerges, where its partial and as yet unconnected primordia can be seen. After all, such are Foucault's efforts—if only we were to look for an example when e.g. examining the discourse on madness (Foucault 1961). They are laborious sets, overburdened with the amount of situations, events, statements, characters, stories, facts. Their complex configuration exposes not only the sense of madness, but especially the dynamics of its image, recorded in these sets, changing its shapes both in time—because it has its own history—and in the space of this abundance scattered habitually in all—including geographic—directions. This means that discourse is better seen not through its own face, but through the conditions in which it appears, and by virtue of watching not the body itself, but its outlines reflected in the matter from which it is made. Like fossilized forms of ancient organisms, moving through time as their spatial negatives and irrefutable pieces of evidence of existence and shape.

In Foucault's description, discourse is non-obvious and unique in nature, which is consistent with the proposed mode of the four principles of his research. Surely, the set, certainly numerous, lively, and intrinsically volatile, yet directly elusive and only indirectly observable, demands unique treatment and non-trivial tools. And this kind of tool is the idea of a complex system, which appears a little early, but its components are already there, ready. These components include, for instance, the idea of a network, already existing as a mature idea in the social sciences and generalized in the following 20 years to make it to the physical world as well, the idea of a system, proposed by Bertalanffy and complemented by others, or the idea of complexity, evolving slowly along the said two, which can also be understood as their speculative culmination. Discourse in Foucault's description seems to fit them well. All that remains is to draw conclusions from this observation, which I will deal with in the final chapter.

In the present story, however, Foucault's insightful portrait of discourse plays the role of an ending, a conclusion in which knowledge (discourse) turns out to be a certain social being, but not necessarily in the sense of Durkheim's classical social fact, but rather one based on a completely different conception of the social, proposed much later by Latour or Luhmann: a chaotic play of numerous, changeable, and moving elements, stabilizing themselves as any social qualification—situations, relations, groups, etc., but only locally and temporarily. And such discourse detaches itself from the human being, understood as the universal condition for the existence of knowledge, which is embodied by discourse in line with the Platonic definition. The shift is not yet final, but quite dramatic: knowledge turns out to be a disposition—not a human one, but an interpersonal one. And immediately when this is accepted, the image of such knowledge explodes, and this knowledge is elevated into areas of complex social processes, where it produces its own increasingly sovereign, independent, autonomous forms of conceptualization of its existence. The most general of those forms brings the concept of the knowledge society, which we addressed in the previous chapter.

However, in addition to this dramatic change at the core, there is an ingredient that makes the whole quite peculiar. An additional circumstance, a seemingly innocent and inherent property, which is the presence of technology increasingly boldly identified with computers and the cyberworld. This convergence has a deep foundation and comes to being as a historical process of understanding human's cognitive efforts, gradually gaining sovereignty over the world they come to learn. It is born as the dilemmas and paradoxes of mathematics and—more specifically—geometry in the nineteenth century, moving quickly into the general field of epistemology, leading as if by the way to the construction of a machine capable of carrying out symbolic inference procedures and, in practice, converting them into calculations. This is where the extraordinary importance of the said digital transformation comes from. Digital technology came to being on the basis of the interpretive and cognitive ideas of this transformation, among which, from this point of view, the most important was the idea proposed by Turing. Its remarkable development, the scale of its applications, and thus its direct impact on reality have obscured its deep sources and imposed a dominant, deterministic approach. This is probably how one would have to define the view of society offered by Masuda, which he referred to as 'computopia'. But the concept of the information society had another creator, who joined and contributed to it in a completely different field and remained outside the Japanese-American competition for primacy as described by Duff. What's more, he merged technique and epistemology into one, clearly drawing on Foucault's inspirations. This creator was Jean-François Lyotard, who wrote a book that has come to be well-known and extremely influential for a very different reason. And this book is *La condition postmoderne. Rapport sur le savoir* (The Postmodern Condition. A Report on Knowledge) (Lyotard 1979).

References

Austin, John Langshaw. 1962. *How to do things with words: The William James lectures delivered at Harvard University in 1955*. Oxford: Clarendon Press.

Bergson, Henri. 1889. *Essai Sur Les Données Immédiates de La Conscience*. Paris: Félix Alcan.

Burgin, Mark. 2015. *Theory of knowledge: Structures and processes*, World scientific series in information studies: Volume 5. Vol. 5. Singapore: World Scientifc Publishing Co. Pte. Ltd.

Deleuze, Gilles. 1988. *Foucault*. Trans. Sean Hand. Minneapolis/London: University of Minnesota Press.

Foucault, Michel. 1961. *Folie et déraison. Histoire de la folie à l'âge classique. Civilisations d'hier et d'aujourd'hui*. Paris: Libraire Plon.

———. 1966. *Les mots et les choses. Une archéologie des sciences humaines*. Paris: Gallimard.

———. 1969. *L'archéologie du savoir*. Paris: Gallimard.

———. 1971. *L'ordre du discours: leçon inaugurale au Collège de France prononcée le 2 décembre 1970*. Paris: Gallimard.

———. 1972. *The archaeology of knowledge and the discourse of language*. New York City: Pantheon Books.

———. 1981. The order of discourse. In *Untying the text: A post-Structuralist reader*, ed. Robert Young, 51–78. Boston: Routledge & Kegan Paul.

———. 2006. *History of madness*, ed. Jean Khalfa. Trans. Jonathan Murphy. London: Routledge.

Hilbert, David. 1899. *Grundlagen der Geometrie*. Leipzig: Verlag von B.G. Teubner. http://archive.org/details/grundlagendergeo00hilb.

Husserl, Edmund. 1996. *Logik und Allgemeine Wissenschaftstheorie: Vorlesungen 1917/18, mit ergänzenden Texten aus der ersten Fassung 1910/11*, ed. Ursula Panzer. Husserliana: Edmund Husserl—Gesammelte Werke. Springer Netherlands. https://www.springer.com/gp/book/9780792337317.

———. 2008. *Introduction to logic and theory of knowledge*. Dordrecht: Springer Netherlands. https://doi.org/10.1007/978-1-4020-6727-3.

———. 2019. *Logic and general theory of science*. Cham: Springer International Publishing. https://doi.org/10.1007/978-3-030-14529-3.

Kirby, Robion C., and Laurence C. Siebenmann. 1977. *Foundational essays on topological manifolds, Smoothings, and triangulations. (AM-88), volume 88*. Princeton: Princeton University Press. https://doi.org/10.1515/9781400881505.

Lee, John M. 2011. *Introduction to topological manifolds. Graduate texts in mathematics*. New York: Springer New York. https://doi.org/10.1007/978-1-4419-7940-7.

Lyotard, Jean-François. 1979. *La condition postmoderne. Rapport sur le savoir*. Paris: Éd. de minuit.

Merquior, J.G. 1985. *Foucault*. Berkeley: University of California Press.

Milkov, Nikolay. 2005. The formal theory of everything: Exploration of Husserl's theory of manifolds. In *Logos of phenomenology and phenomenology of the logos. Book one: Phenomenology as the critique of reason in contemporary criticism and interpretation*, ed. Anna-Teresa Tymieniecka, 119–135. Dordrecht: Springer Netherlands. https://doi.org/10.1007/1-4020-3680-9_7.

Quine, Wilard V. 1960. *Word and object*. Cambridge, MA: MIT Press.

Riemann, Bernhard. 2016. *On the hypotheses which lie at the bases of geometry*, ed. Jürgen Jost, Classic texts in the sciences. Cham: Springer International Publishing. https://doi.org/10.1007/978-3-319-26042-6.

Rorty, Richard M., ed. 1992. *The linguistic turn: Essays in philosophical method*. Reprint ed. Chicago: University of Chicago Press.

Smith, David Woodruff. 2002. Mathematical form in the world. *Philosophia Mathematica* 10 (2): 102–129. https://doi.org/10.1093/philmat/10.2.102.

Torretti, Roberto. 1978. *Philosophy of geometry from Riemann to Poincaré*. Dordrecht: Springer Netherlands. https://doi.org/10.1007/978-94-009-9909-1.

Tu, Loring W. 2011. *An introduction to manifolds. Universitext*. New York: Springer New York. https://doi.org/10.1007/978-1-4419-7400-6.

Wittgenstein, Ludwig. 1958. *Philosophical investigations*. 2nd ed. Oxford: Basil Blackwell.

Chapter 8
Lyotard and Applied Knowledge

Jean-François Lyotard, a prominent French philosopher, is known as the one who taught everyone the meaning of a concept that continues to inspire fear among academics longing for good old science and among everyone longing for the good old world where truth had its place, just like goodness or beauty. That concept is postmodernism. The book that acted as the first guide, so to speak, as well as a kind of manifesto on postmodernism, was the already mentioned *La condition postmoderne. Rapport sur le savoir* (Lyotard 1979) (*The Postmodern Condition: A Report on Knowledge* (Lyotard 1989)). A widely commented and influential work, it was written as a report commissioned by the President of the Council of Universities at the Québec government. Perhaps this utilitarian and practical reason for its coming to being explains the unusual ambivalence of the book, written in language that is firm, that makes the author's theses clear from the beginning. But it is also a language that opens up a serious, deep, and multifaceted reflection on the state of societies at the end of the twentieth century. The latter circumstance has made the book the source and cause of a great number of disputes and commentaries. Here, however, we will focus on its remarkable boldness, its daring nature, possible thanks to and aided by a consistent and coherent line of thought.

In the introduction, which lays out the crux of the lecture with astonishing simplicity and sets up the conceptual figures on the playing field, Lyotard refers to two entities—to make it clear that he intends to explore their mutual configuration and reveal the far-reaching consequences of their intertwining. These entities are knowledge and society. The position they take against each other is not a simple arrangement, and the author puts a lot of effort into explaining them. Chapter 1 begins with a hypothesis that reads: "Our working hypothesis is that the status of knowledge is altered as societies enter what is known as the postindustrial age and cultures enter what is known as the postmodern age" (Lyotard 1989, 3). This hypothesis describes the interrelation of knowledge with social and cultural processes, but does not completely determine its direction, the potential mutual influence of one on the other. For there is no such determination either. Instead, there is a condition, the

R. Maciag, *Knowledge as a Tale*, https://doi.org/10.1007/978-3-031-69820-0_8

theoretical basis of which Lyotard presents only later, in the last two chapters, which offer a lecture on a kind of descriptive and cognitive paradigm. The idea of this condition, this state, is a great work of art. A work of art that is truly impressive given its visionary nature and the backdrop of the times in which it was created, maintaining its relevance to this day. A work of art that has not received any special attention in the literature either. This is certainly the case because Lyotard makes use of ideas that emerged in advanced mathematics and physics, thus proving—incidentally—the utter failure of the naive division of sciences into exact sciences and the humanities, which sit right next to each other after all. As if there was any point in separating them at all, since they intermingle and intertwine in the general universe of knowledge and science.

Lyotard also exposes the weakness of distinguishing between scientific knowledge and common knowledge, which becomes apparent at the level of his reflection on knowledge as such—that is, the level of metaknowledge, knowledge of knowledge, which is not only possible, but even essential to the understanding of the processes of cognition. In doing so, he refers to Gödel, as well as to the idea of an axiomatic system developed separately by Giuseppe Peano and David Hilbert in the late nineteenth century. Although their names are not mentioned, there is a clear reference to their concept, which triggered a profound upheaval concerning mathematics as a tool of cognition. Kline describes it as follows: "The two-thousand-year-old conviction that mathematics was the truth about nature was shattered. But the mathematical theories now recognized to be arbitrary had nevertheless proved useful in the study of nature. Though existing theories historically owe much to suggestions from nature, perhaps new theories constructed solely by the mind might also prove useful in the representation of nature. Mathematicians should then feel free to create arbitrary structures" (Kline 1990, 1036). These 'arbitrary structures', which are purely the work of the mind and highly discretionary, were not based on observing nature and capturing these observations in formal models, but offered a completely reverse, opposite form—seemingly unreasonable at first glance. According to this form, models were built as free creations, as results of imagination and nothing more. Only in the next step was their possible usefulness for describing the world verified. And even this was not necessarily the case, because this adherence to reality was not the key to making sense of these structures. This key was the consistency of the invented systems of principles—axioms, deriving their power of being solely from mutual compatibility, which was otherwise not all that easy to achieve. And it was a new experience, because the 'old' theories, referring to the world, turned out to be consistent as if by default: the world, after all, seems consistent.[1]

This mathematical step, which Lyotard dissects into its various continuations, referring to Mandelbrot's fractals, Thom's catastrophe theory or quantum physics, is obviously not just a local event. It poses a fundamental question about the source

[1] Although concerns about this condition date back to very old times. They are deposited, for example, in the cosmogonies of all religions and beliefs, intent to prove the pre-established harmony.

of theory and offers a perverse—though otherwise dramatically accurate and general—answer: it is the human mind (Lyotard adds another element to this order—language, but we shall deal with that in a moment). Mathematics breaks free, abandoning the restraining shackles of cognition to represent reality and opting for pure speculativeness, which turns out to be as useful in representing the world as strenuous observations and the conclusions that flow from them. Kline writes: "Since the axioms were no longer truths about the concepts involved in them, the physical meaning of these concepts no longer mattered. This meaning could at best be a heuristic guide when the axioms bore some relation to reality. Thus even the concepts were severed from the physical world" (Kline 1990, 1035).

Lyotard consistently draws conclusions from this procedure and extends it to encompass all knowledge. He is not the first to do this. A similar line of thought, though much more speculative, is offered by Popper (1935), who analyses the ability of axiomatic systems to represent reality in detail. One can figure that they are the inspiration for his introductory statement, concerning this extraordinary moment when a concept that may later become a scientific discovery appears in the world. It has every right to be called unusual, special, because it is the result of unlimited freedom of creation and can come from anywhere: "what you will" as Popper claims directly (Popper 2002, 9). Lyotard pays special attention to the same aspect that underlies the new thinking about mathematics. It is about bringing to the fore not the object of knowledge (theory) itself—meaning not, for instance, a ready-made mathematical system or a particular geometry (many such can arise from different assumptions), but the path that leads to them. The way that leads to such a state is to realize the idea of free assumptions—axioms, speaking precisely. However, making them the object of consideration results in a situation where it is not so much the work as the conditions of its creation that turn out to be important—as important as the outcome. Such reasoning leads Lyotard to his main conclusion regarding the legitimacy of knowledge: "When a denotative statement is declared true, there is a presupposition that the axiomatic system within which it is decidable and demonstrable has already been formulated, that it is known to the interlocutors, and that they have accepted that it is as formally satisfactory as possible" (Lyotard 1989, 43).

Knowledge is thus a kind of agreement based on mutually accepted assumptions. When these assumptions change, new knowledge must be created. This approach leads to another firm conclusion regarding a certain necessary cognitive context: "the striking feature of postmodern scientific knowledge is that the discourse on the rules that validate it is (explicitly) immanent to it" (Lyotard 1989, 54). Lyotard defines discourse somewhat differently than Focault, although there is no doubt that, as he writes, "scientific knowledge is a kind of discourse" (Lyotard 1989, 3). In doing so, he places the phenomenon in a very specific language environment, which we will return to later. Discourse understood as the context and legitimization of science draws attention to the deep and dramatic social role of science. This concept of discourse also opens the way to justifying the notion of modernity and then postmodernity understood as something that comes *after* modernity. The principle of the former is to refer to the discourse that legitimizes science as valid and without alternatives. It can change, though, and Lyotard lists several possibilities that hint at various philosophical sources of such legitimization: the dialectic of Spirit, the hermeneutics of meaning, the emancipation of the rational or working subject, the creation of wealth. These metanarratives, as Lyotard calls them, not only make it possible to construct philosophical assumptions about knowledge and the world revealed through that knowledge, but are also the basis for their social—e.g. institutional—implementation. They thus lead to a certain social model and its ethical assumptions—concerning, for example, justice. As a result, postmodernism, the era that came at the end of the twentieth century, breaks the principle of the primacy and universality of metanarratives; metanarratives lose our trust, become questioned and represented in equal number.

Lyotard's approach, based strongly on a deep rethinking of the principles of cognition at the root of mathematics and geometry, i.e. the sciences that have traditionally provided a strong basis for the credibility of the natural sciences, must have caused common concern—and continues to do so, in fact. For it touches the very foundations of the way we understand our knowledge of the world, and it does so by painfully stripping us of the dream of ultimate, complete, objective or true knowledge. It makes us realize that knowledge is entangled with the conditions of its formation, and this entanglement is deep, immanent—and cannot be circumvented. Lyotard takes this observation to the level of social practice, for it is the best context to consider the conditions of a technical agreement a model of which is axiomatic systems. At this level, said observation turns into two shocking questions: "who decides what knowledge is, and who knows what needs to be decided?" (Lyotard 1989, 9), which immediately draws our attention to a strongly political context, or—to put it simply—to power relations, thus opening the field of critical social reflection.

According to the logic of the discourse just described: "the relation to knowledge is not articulated in terms of the realization of the life of the spirit or the emancipation of humanity, but in terms of the users of a complex conceptual and material machinery and those who benefit from its performance capabilities. They have at their disposal no metalanguage or metanarrative in which to formulate the final goal and correct use of that machinery. But they do have brainstorming to improve its performance" (Lyotard 1989, 52). Paying attention to the pragmatic, goal-oriented treatment of knowledge involves a deeper—or even more practical—examination of the social mechanism of its functioning, but the path that ends with pure technology leads Lyotard through language as the domain of discourse. Lyotard also takes this opportunity to equate scientific and non-scientific knowledge, which must happen, because their ontological basis is common. And this basis is discourse: "drawing a parallel between science and nonscientific (narrative) knowledge helps us understand, or at least sense, that the former's existence is no more-and no less-necessary than the latter's. Both are composed of sets of statements; the statements are 'moves' made by the players within the framework of generally applicable rules; these rules are specific to each particular kind of knowledge, and the 'moves' judged to be 'good' in one cannot be of the same type as those judged 'good' in another, unless it happens that way by chance" (Lyotard 1989, 26).

Speaking about the world is a necessary cognitive act. This happens regardless of formal circumstances, and one to claim that there are formalisms able to resist this speaking would be by all means wrong. In particular, it is necessary to get rid of any illusions in the case of science: "science is a model of an 'open system', in which a statement becomes relevant if it 'generates ideas', that is, if it generates other statements and other game rules. Science possesses no general metalanguage in which all other languages can be transcribed and evaluated" (Lyotard 1989, 64). Thus, science is linguistically dispersed in the sense that language is understood here simultaneously as a set of rules of speech, which is subject to the logic of axiomatic systems: free and undetermined. Languages tend to proliferate. No single one of them can claim supremacy. Thus, a space of equivalent statements is generated. These statements are discourses, which are dynamic, changeable. But the same reason, their multiplicity that makes them unable to reign supreme, brings them closer to other narratives—colloquial, fairy-tale-like, etc. This is another version of the social constructivism in the bosom of science that Knorr-Cetina, Woolgar or Latour wrote about in a similar period (Knorr-Cetina 1981; Latour and Woolgar 1979). The world of knowledge thus outlined also demands a new and ethically mature attitude and regulation, accepting—and even affirming—this new world, the basis of which should be the recognition that "language games are heteromorphous" (Lyotard 1989, 66).

Latour borrows the notion of language games directly from Wittgenstein, from his *Philiosophical Investigations* (Wittgenstein 1958), citing one well-known passage in which the latter compares language to an old city surrounded by new neighborhoods, which is a gradually thickening labyrinth with no clear boundary, as if dissolving into the landscape. This passage has already been mentioned, but its fitness to the context make it all the more reasonable to quote it: "Our language can be seen as an ancient city: a maze of little streets and squares, of old and new houses, and of houses with additions from various periods; and this surrounded by a multitude of new borough with straight regular streets and uniform houses" (Wittgenstein 1958, 8). An apotheosis of landscape, reassuring and carrying the fear of loss at the same time. This complex image is expanded a few sentences later: "to imagine a language means to imagine a form of life" (Wittgenstein 1958, 8), adding to the essential density, significance, and—at the same time—tension that the author manages to include in this vision.

As we remember, Foucault had no doubt that the act of speaking is dialectically simultaneously free and deeply regulated, for the control of the content and manner of speech is an emanation of pure power. In the case of Lyotard, who wrote his *Condition* almost a decade after the synthetic lecture on the subject that Foucault gave at the Sorbonne in 1970, the anxiety is similar, but much more tangible, concretized. Lyotard deals with similar dilemmas as Foucault, and the two are often mentioned together as exponents of French poststructuralism, along with Derrida and others. And this is obviously unfair, given the originality of each of the mentioned and unmentioned philosophers listed together in this current. Leaving others aside, however, it should be recalled that it was Lyotard who combined the social and historical order of the rise and fall of the great projects of knowledge with a purely informational context, becoming an advocate of the idea of—actually—a digital society, which is clear in his *Condition*. And this happened relatively early. Earlier, for example, than the emergence of Masuda's idea, which was discussed at the beginning of Chap. 3.

The last two steps of Lyotard's reasoning described tie this reasoning together in the sense that knowledge and society—the two entities we began with—turn out to be necessarily intertwined and indispensable to each other. The first step has already been taken: that of presenting language as a proper domain of knowledge. The second involves Lyotard's risky proposal to link it to the field of information technology, making the philosopher one of the most important and—at the same time—lesser-known theorists of the so-called information society. Lyotard has no doubt when it comes to this relationship. When he defines knowledge as discourse, he immediately adds: "it is fair to say that for the last forty years the 'leading' sciences sciences and technologies have had to do with language" (Lyotard 1989, 3). He then goes on to list eight examples, which include the following five: problems of communication and cybernetics, modern theories of algebra and informatics, computers and their languages, problems of information storage and data banks, telematics and the perfection of intelligent terminals. When it comes to the first of the said examples, he refers to Norbert Wiener, the founder of cybernetics. In the case of the second—to John von Neumann, the creator of the machine that is the prototype of all modern computers.[2] The problem of storage memory and data banks includes a reference to Simon Nora and Alain Minc's report on the computerization of society, written at the request of French President Valéry Giscard d'Estaing (Nora and Minc 1978).

The connection between the aforementioned technologies and language seems surprising at first glance, but Lyotard defines this connection in a very simple, even trivial way, and writes that the aforementioned technological transformations will have an impact on two fundamental functions of knowledge: research and transmission[3] (Lyotard 1989, 4). Here, in the context of these technologies, language is reduced to communication, which turns out to be opaque, vague, but strongly determined by said technologies. Lyotard considers the development of information technology as rapid and inevitable, but he describes the impact of this development on language as the building block of knowledge somewhat inaccurately. Actually, he understands it quite superficially, speaking of translating the content of knowledge into bits or 'machine language', which can limit or modify this content. Fortunately, the metaphorical nature of this description, especially in the second case, opens up some possibilities for expanding the purely communicative function of language as the only sphere subject to the impact of technology. Although Lyotard allows for another kind of link between knowledge and technology, in which the latter plays the role of a certain structural pattern, a morphogenetic source, as

[2] I discuss this reference, von Neumann's role, his famous *First Draft*, where he presented the design principles of the computer, in my book *Deus ex machina. Nowe media i ich projekt poznawczy* (Maciąg 2012).

[3] Lyotard's term 'connaissances' is translated as 'learning' e.g. (Lyotard 1989, 4).

exemplified by genetics making use of the constructs of cybernetics, he completely fails to elaborate on this argument.

Meanwhile, the interplay between knowledge and information technology, which will take the form of digital technologies, can be deep and intense. For instance, the understanding of knowledge under the influence of these technologies is completely changing when it comes to the internal solutions of the latter. Knowledge becomes a technical factor, a kind of resource subject to algorithmic logic. But the emergence and development of this relationship can also be understood as a story about the production of a specific model of knowledge, and since knowledge means a certain image of the world—epistemology is closely linked to ontology, the world turns out to be different and hitherto unknown. The concept of digital transformation can be interpreted in this very way, rejecting the perception of this transformation as a solely pragmatic and trivial change in social relations—mainly economic ones—as a result of the development and practical use of said technologies. What needs to be stressed is that the origins of this thinking are also rooted in the same mathematical and geometrical premises Lyotard mentions. The breakthrough brought about by the possibility of creating uninterpreted axiomatic systems (i.e. not represented in the real world) implies freedom and imagination in the creation of premises and models, reduces science to a series of instantaneous and local paradigms, and even leads to a kind of cognitive anarchy. This breakthrough occurs first in the realm of geometry, then proceeds to the field of mathematics, triggering the need for an overarching metatheoretical inquiry. It is within the domain of metamathematics where the idea of the so-called Turing machine—or conceptual model of a computer—is born. But this is just a kind of side effect in the process of reconstruction of knowledge itself. The intuitions and findings concerning such a reconstruction can be found in Lyotard's *The Postmodern Condition*.

Lyotard predicts the logic of transformation of knowledge due to computerization very precisely and accurately. The first consequence is the gradual exclusion of a human as an essential participant in the processes of this transformation, which marks a significant change from the classical approach, dating back to Plato, as described in Chap. 4. Lyotard writes explicitly that "we may thus expect a thorough exteriorization of knowledge with respect to the 'knower'. at whatever point he or she may occupy in the knowledge process. The old principle that the acquisition of knowledge is indissociable from the training (Bildung) of minds, or even of individuals, is becoming obsolete and will become ever more so" (Lyotard 1989, 4). The German term *Bildung*, meaning education, refers to an idea inspired by Hegel, present in the classical education advocated by Goethe, Schiller, and others (Borowski 2009), and introduced by Wilhelm von Humboldt. It involved the 'forging' of both wise and righteous personalities as an essential component of university education. This idea cannot be fulfilled if knowledge transfer is reduced to a technical process. However, Lyotard's acute—and legitimate—observation is at the very beginning of the changes later experienced by knowledge, understood as an element of human subjectivity. Ethical or psychological issues start with a much deeper process, which involves breaking the subjective nature of knowledge in general, which appears in the context of advanced technologies—like in the case of e.g.

so-called artificial intelligence. In the 1970s, however, digital technologies are at a very early stage of their development, and it's difficult to predict their further reach. Yet, Lyotard's 'hunch' regarding this field proves to be as correct and valid as possible.

From the process of peculiar dehumanization of knowledge, Lyotard derives the effect of its objectification, or even reification, which then immediately turns into an ordinary process of so-called commodification. This notion, taken from the brute dictionary of management, unknown to Lyotard, means reducing a thing to a commodity, a product, a good that is traded. The change in the ontological status of knowledge is to make it more real and at the same time give it an economic context. Lyotard draws here directly from Marx's analyses, modifying them slightly in the process. The latter perceives knowledge through its products, such as machines; Lyotard grants knowledge similar productive power, while emphasizing its immanence and emancipation. He writes: "knowledge is and will be produced in order to be sold, it is and will be consumed in order to be valorized in a new production: in both cases, the goal is exchange" (Lyotard 1989, 4). Knowledge detaches itself from any of its administrators. In particular, it detaches itself from man, ceasing to be their exclusive property. Since knowledge boils down to information in the new post-modern world, it is information that becomes the direct and technically graspable, real element of exchange. A traded good. This process is so intense that information creates flow structures similar to money, becoming a means of payment or a means of accumulating wealth, making the picture of a society built according to its logic complete.

Lyotard reinforced his description of information and the consequences of placing it at the heart of social processes with what he found in a report that generated considerable interest when it came out. This report is a document written at the express request of French President Valéry Giscard d'Estaing, published in January 1978. Its main author—and recipient of the request—was a high official of the French government, Simon Nora, who chose Alain Minc to assist him in the task. The result of their collaboration was a comprehensive analysis of the state of information technology, which was relatively advanced in France at the time. It was published as *L'informatisation de la Société* (Nora and Minc 1978). The work became popular and two years later was translated into English and published with a foreword by Daniel Bell, entitled *The Computerization of Society: A Report to the President of France* (Nora and Minc 1980a). The authors focused on the phenomenon of data and data sets as one of the most important effects of technical information processing, highlighting the emerging and future processes related to the existence of this phenomenon. They argued, for example, that "information is inseparable from its organization and form of storage. In the long term, it is not only the advantage that the knowledge of one or other piece of data provides. The knowledge will end up by modelling itself – as it always has done – on the stocks of information" (Nora and Minc 1980b, 28), which was, on the one hand, very practical, as expected of the work of officials, but on the other hand, it was surprising in its revealing simplicity and bluntness of insights.

Nora and Minc's report also referred to the knowledge they perceived through information. They believed that the rapid development and transformation of information management would change the conditions of the functioning of knowledge, which would then lead to profound changes in the functioning of society. They wrote, for example, that "the establishment of data banks will be the beginning of a rapid restructuring of knowledge" (Nora and Minc 1980b, 30). It is important to do justice to the authors of the report and add that despite reaching to a seemingly administrative source of inspiration, their conclusions were deep and insightful, and they were able to evaluate the effects of informatization also in other contexts than the purely social one. This made it possible for them to offer quite broad, far-reaching judgments, such as this one: "data processing is introducing a phenomenon comparable to that introduced by the appearance of writing" (Nora and Minc 1980b, 29). They were also able to interpret such a diagnosis analytically, using, for instance, the concept of culture: "the cultural model of a society also depends on its memory, control of which largely conditions the hierarchy of power. Access to infinitely greater sources of information will entail basic changes and will affect the social structure by modifying the procedures for acquiring knowledge" (Nora and Minc 1980b, 30).

Lyotard refers to Nora and Minc's report explicitly several times, but it is quite easy to see how deeply inspired he was by that publication in general. This is especially true in the case of the link between knowledge and information, which becomes the main auxiliary phenomenon he analyzes in the context of knowledge. This configuration perfectly fulfilled the objectives of the design engineered and provided by Nora and Minc. Lyotard clearly follows the line of thought they opened, expanding it and taking it further. At the same time, he is much more forceful, more resolute, and even beyond that: he introduces a strong and unambiguous social context, for example, by setting his analysis in a Marxian perspective, which is not annoying, bothersome, but explicit, which was, of course, something French government officials could not afford to do in a similarly open manner. The conclusions that Lyotard draws thus go further and are more dramatic. They also become a guideline for political regulations, which remain relevant till this day, when he argues that computerization "could become the 'dream' instrument for controlling and regulating the market system, extended to include knowledge itself and governed exclusively by the performativity principle. In that case, it would inevitably involve the use of terror. But it could also aid groups discussing metaprescriptives by supplying them with the information they usually lack for making knowledgeable decisions. The line to follow for computerization to take the second of these two paths is, in principle, quite simple: give the public free access to the memory and data banks." (Lyotard 1989, 67).

Lyotard does not display the same analytical insight as Foucault when he writes about the discursive nature of knowledge, and discourse is not an issue he is deeply interested in. It could be fair to assume that he considers the job done. He is heading—in line with his task, in fact, as he is writing a commissioned report—at a certain social project, understood quite literally. Webster, in analyzing theories of the information society, describes Lyotard's contribution and pigeonholes him under the label 'postmodernism', putting him in the company of such figures as Mark Poster, Gianni Vatimo, and Jean Baudrillard (Webster 2006). At the same time, he still highlights his uniqueness, which is manifested in the French philosopher's focus on the intriguing and clearly fundamental linkage between information and knowledge, pushing this relationship in a very practical direction. A new axis of social processes is being generated, governed by a very simple principle. Information and knowledge "are produced more and more often only where they can be justified by efficiency and effectiveness" (Webster 2006, 252). Lyotard refers to this area as *la performativité* (Lyotard 1979), which is translated as 'performativity'. This performativity cannot be understood directly, but is defined within the game that knowledge plays with its institutional, political context. Its ultimate stake is the possession of knowledge, and since the value of the latter can be measured by money, knowledge ultimately finds itself in the realm of economic entities—as a valuable resource, and political entities—as a factor of advantage.

Lyotard's reasoning has a strong Marxian basis—as Webster does not fail to emphasize. It also seems as if too simple, since the relationships between knowledge and power are, after all, well known. However, Lyotard analyzes social and political mechanisms using completely new contexts: a technological one, an informational one, and a digital one, which produce said relationships as if anew. This makes the story that began as the tale of Masuda and his computer utopia, wandering through the ideas of Machlup and Bell all the way to Lyotard complete. This is a story about knowledge, whose new role has been described by Bell in a quite calm and composed manner: "the major source of structural change in society – the changes in the modes of innovation in the relation of science to technology and public policy – is the change in the character of knowledge: the exponential growth and branching of science, the rise of a new intellectual technology, the creation of systematic research through Research and Development budgets and, as the calyx of all this, the codification of theoretical knowledge" (Bell 1973, 44). In the case of Lyotard's left-wing and critical mindset, it doesn't look so good anymore. However, both have no doubt that deep and powerful processes have already been set in motion. In light of the above, Masuda's enthusiasm seems naively childlike, carefree. After all, we could sense this naivety earlier, in his clear and orderly descriptive structures.

The future has added more specific scenarios for using knowledge, as well as for increasing its effectiveness. Data banks, for instance, have indeed turned out to be a purely commercial and political tool, taking on the commercial name of *big data* (Chandler and Fuchs 2019). This phenomenon is part of an extensive story about such a variant of knowledge implementation that is emerging in connection with digital technology, and although its history far predates Lyotard's text discusses here, Lyotard does not mention it at all. This is about the models of knowledge that appear, for example, in connection with the development of so-called artificial intelligence—especially at a time of the growing popularity of the symbolic approach, whose name comes precisely from the mode of representation of knowledge it refers to. It can be seen gaining momentum extensively in the late 1960s and early 1970s. At the same time, it is the beginning of research and technical solutions whose rapid advancement continues to this day, although the symbolist approach has given way to other ideas. These 'other ideas' include first and foremost the so-called connectionist solutions, i.e. artificial neural networks, which today have proven to be the most common version of artificial intelligence. There are also two other projects that are direct manifestations of the utilitarian nature of knowledge: the purely pragmatic Knowledge Management (KM) and the rational Knowledge Organization (KO). The former makes knowledge simply a resource that can be consumed similarly to other resources disposed of by an organization (an enterprise). The latter shows the environment of knowledge and information as a territory for reasonable, efficient, and safe exploration. Both view knowledge exclusively from a practical point of view. It is reasonable to take a moment now to consider them in more detail in order to get a good perspective and sound judgment of the development of the phenomenon of knowledge.

Knowledge Management

The explorations into knowledge management begin with a textbook on the subject, written by Kimiz Dalkir, published in 2005 (Dalkir 2005). The 'institution' of the textbook itself is quite unusual. A textbook cannot be controversial and must be comprehensive, which already puts it in an absurd position, given that it is one of the most popular emanations of knowledge. It is also, of course, full of compromises, which is the only way to deal with the plethora of existing paradoxes. One such paradox is, for instance, that a textbook should tell the truth, but in the end it offers at most a temporary agreement on the commonly adopted beliefs. Writing a textbook must be enjoyable: it is an experience in building a lasting and reliable picture of the world. But it is also a never-ending failure when it lines up with a long line of textbooks that are already outdated, a reminder that they are doomed to the same fate eventually. It's not easy to accept these pitfalls and write a textbook given the circumstances. The textbook we're dealing with here is a textbook on knowledge management, so it's actually autothematic, but its author doesn't emphasize this fact, because such a perspective by itself can give rise to concern. It's a textbook that is completely aware of the significance of its subject and firmly sets rhetorical figures, concepts, and structures on the playing field.

The energy comes from certainty of the importance of knowledge combined with an enthusiastic conviction of its availability. Knowledge management provides the key to this asset. This belief emerges from a multitude of variants and variations. Dalkir reports that he encountered more than a hundred definitions of knowledge management, of which he considers at least 72 to be very good (Dalkir 2005, 4). It appears, therefore, that definitions do not strive for ultimate unification, to capture a *definiendum* that is essential, in spite of the obvious and constantly experienced fact that this is simply not possible. A definition, after all, chooses its object from the variety provided by the world—and makes it unique by endowing it with an identity. But it can also be the other way around. The definitions of knowledge management are of the opposite nature. Their value lies in the diversity that comes from the vastness of the area they encompass. Dalkir gives a few examples, which quickly grow into an impressive number of different kinds of approaches, embedded in all kinds of research and scientific fields. These include "organizational science, cognitive science, linguistics and computational linguistics, information technologies such as knowledge-based systems, document and information management, electronic performance support systems, and database technologies, information and library science, technical writing and journalism, anthropology and sociology, education and training, storytelling and communication studies, collaborative technologies such as Computer-Supported Collaborative Work (CSCW) and groupware as well as intranets, extranets, portals, and other web technologies" (Dalkir 2005, 8). Knowledge management proves to be truly multidisciplinary, which is, as one might expect, due to the very nature of knowledge, which can encompass every aspect, every bit of the world. That's why definitions derive their strength from diversity rather than precision or accuracy—they are self-reflexive. Knowledge about knowledge management is, by necessity, knowledge about knowledge.

Dalkir finds the diversity and richness of knowledge management somewhat concerning, which makes him look for good features that distinguish this type of management from any other knowledge-seeking situation—i.e. from the ubiquitous activity and process, which ultimately manifests itself by pointing to two competing and complementary fields of knowledge: the so-called *tacit knowledge*, spoken of by Michael Polanyi, and *explicit knowledge*. According to Dalkir, the former is all the more valuable the more difficult it is to articulate, and this seems to make it fall into a paradox, or at least make the question of how it exists burning. It resides "within the heads of *knowers*" (Dalkir 2005, 10), Dalkir writes, and includes various forms of expert knowledge, know-how or know-why (an allusion to Ryle's ideas [Ryle 1949]), is subjective, and often refers to beliefs or experiences. Such knowledge cannot be made public, and this is not the goal of knowledge management. Davis and others, citing Goldman, use slightly different terminology and a different tradition in trying to define it: "from an epistemological standpoint, most of what is characterized as organizational. knowledge falls within the purview of "weak knowledge"" (Davis et al. 2005, 5). Such knowledge fulfills only part of the Plato's definition—it is a true belief, but it no longer has to be 'captured in strict terms', which in the English translation is referred to as *justified.*[4] It derives its power from consistency within a broader perspective and a pragmatism that binds and justifies it through action, experience or experimentation. The latter aspect was emphasized by Nonaka and Takeuchi, who can be considered the fathers of knowledge management (Nonaka and Takeuchi 1991).

Kahrens and Früauff, critically analyzing Nonaka's concept of knowledge creation in organizations, remind us of the three basic forms of knowledge, which were noticed and introduced gradually: the first—earliest—was *tacit knowledge*. The second is described as follows: "knowledge is also the actuality of skilful action

[4] I discuss this issue in more detail in Chap. 1.

and/or the potentiality of creating situations to permit (skilful) action" (Kahrens and Früauff 2018, 54). The third considers knowledge as a certain continuum in which its tacit and explicit variants merge into a whole. These forms and, at the same time, spaces of knowledge retain and follow the fundamental division into tacit and explicit knowledge, and only then grow into numerous detailed, more specific variants. It seems further that the reflection on knowledge in organizations still serves the two basic goals that Wiig defined in the late 1990s, a period of rapid development of this reflection. They are as follows: first, to make an enterprise function as intelligently, as smartly as possible, and, second, to take advantage of the best properties and qualities of the knowledge resources in possession (Wiig 1997). This is also how Dalkir ultimately sums up the essence of knowledge management in his textbook: "Knowledge management represents a deliberate and systematic approach to ensure the full utilization of the organization's knowledge base, coupled with the potential of individual skills, competencies, thoughts, innovations, and ideas to create a more efficient and effective organization" (Dalkir 2005, 3).

The atomic pragmatism of this approach is striking, especially in the context of Polanyi's concept. After all, he built the idea of tacit knowledge on a very extensive, virtually metaphysical foundation. The latter property appears explicitly, as something straightforward, given the fact that he supports his construct with an ontological justification (Polanyi 1966, 13). Such panache is obvious in light of Polanyi's task, which is to establish a theory of the foundations of the structure of the world as we know it. This structure features an order in which living beings are the primary pretext for the production of its sense: "But dead matter, matter that is both lifeless and deathless, takes on meaning by originating living things" (Polanyi 1966, 91). It is the process of life—as a separate but powerful and decisive causal factor—that paints the visionary picture of the development of the cosmos, in which meaning is the essence of life. This vision reaches biblical proportions: "The field of new potential meanings was so rich that this enterprise, once started, swept on toward an infinite range of higher meanings, unceasingly pouring them into existence, for the better part of a billion years" (Polanyi 1966, 91). In this general and ultimate process, the concept of meaning seems closer to the idea of the logos, referred to already at the end of Chap. 2, signifying immanent, timeless, and ultimately extra-human meaning. This notion brings us closer to knowledge, which exists, as Polanyi argues, as a separate, silent, *tacit* dimension of the world. Thus, tacit knowledge is a component of the act of establishing, forming the world, and the practical effect looks only like a constituent element, some necessary and secondary consequence of the wonderful harmony that connects man and the world: "My search has led me to a novel idea of human knowledge from which a harmonious view of thought and existence, rooted in the universe, seems to emerge" (Polanyi 1966, 4).

Knowledge management, which grew exponentially as a field in the 1990s and which, fortunately for historiographers, Wiig covers in great detail and highly skillfully in the aforementioned article, acts as a certain simplification of Polanyi's ideas. The process he referred to as 'the cosmic emergence of meaning' (Polanyi 1966, 92) has shrunk to a certain utilitarian strategy, although it has not been completely stripped of its deep, unfathomed, hidden nature. This can be seen perfectly in the consequences that arise when the idea of tacit knowledge is introduced into the field of reflection and examined in various interpretive configurations. We can see it clearly in the descriptions of knowledge management that Syed et al. collect in pursuit of their ambitious goal to create and offer a complete and comprehensive overview (Syed et al. 2018). These descriptions make it possible to derive a synthetic argument, difficult to achieve in the context of the multitude of insights presented, bringing one area to the fore: technology, and within it—information technology. For this reason, the future of knowledge management "is likely to focus on connectivity, collaboration and co-creation, thus integrating the information and socialising aspects of managing knowledge" (Hislop et al. 2018, 696).

This conclusion, like a strong beam of light, points to the main theater of events, but it is also general and basic enough to recall Polanyi's panache. It also makes it impossible to confine knowledge management in a non-ambiguous and simple field of inquiry. It explodes suddenly and rises to a new, higher level of generalization. It is more modest, does not feature a 'cosmic' dimension, and does not necessarily refer to a life-based world order, but upholds the claim of the specificity of the knowledge on which it is actually based—the very one that Polanyi discovered. At the same time, it takes knowledge to a similar level of generality, although describes it with a completely different, much more precise and disciplined conceptual apparatus, which can be called 'social'—in the broad sense given to the word by the reflection on the presence of man in the world as a collective rather than as a single individual. It had already developed rapidly after the publication of Polanyi's book, whose author could not use its tools and was guided by hunches expressed rather in a language borrowed from various philosophical and perhaps even esoteric traditions. The arrival of computers in this context should be considered an expected necessity.

But let's face it. Polanyi's transcendental vision is discarded and replaced by proper engineering, allowing for the use of layers of knowledge in an orderly and efficient manner. This is the purpose of the methods the list of which starts with Nonaka's (1991; Nonaka and Takeuchi 1991) classic concept called SECI (Socialization, Externalization, Combination and Internalization), constructed, as Kahrens and Früauff claim, as a model representing "a knowledge creation process as played out through a continuous dialogue between tacit and explicit knowledge" (Kahrens and Früauff 2018, 53). The approach is technical, accompanied by the idea of a cosmic emergence of meanings accessed as if directly and remaining forever hidden in the mystery of this direct insight. It turns into practical familiarity—a skill that can be taken advantage of, although the world at this level turns out to be a very rich deposit, soon becoming an object of greedy exploitation.

Knowledge Organization

A somewhat different idea of interpreting knowledge as a certain materialized and exploitable world is an idea that is less obvious than knowledge management. Its back story, too, contains a certain element of tragedy that gives it its meaning and reveals the stakes. We are talking about an area known as Knowledge Organization (KO). Its main theorist is Birger Hjørland, thanks to whom this modest project was reborn as an ambitious undertaking. In the current canonical description of it, available on the official website of the International Society for Knowledge Organization,[5] its source appears first among the main problematic fields associated with it. It is the library, or more precisely: the logic of its world, based on a strict order of catalogs with layers of knowledge behind them. It's a very specific approach. At first glance, it appears as purely technical, because it exists out of necessity, as a response to the growing piles of increasingly dusty books. But it's also a world—one that is the everyday of knowledge. It is its current state of existence, in which practicism touches metaphysics directly, as it is an ever-expanding direct illustration of the totality of man's epistemic efforts.

The moment when the cluster term 'knowledge organization' was created is described by Hider (2018). He highlights the specific organizational, technical, and historical circumstances: "The term 'knowledge organization' was chosen for the English name to represent wider interests than classification, although these did not at first extend to other IO [Information Organization – author's note] activities such as descriptive cataloguing: it was abstract rather than recorded knowledge that was to be organized into schemes and vocabularies" (Hider 2018). However, the field is constantly being torn in two different directions, concerning understanding knowledge in different ways—one narrower, concerned with library practice, the other broader, aspiring to general and abstract knowledge constructs. Such is the ultimate definition offered by Hjørland in two parts, so to speak: "KO is about describing, representing, filing and organizing documents and document representations as well as subjects and concepts both by humans and by computer programs" (Hjørland 2016). He gives much attention to the two said notions, i.e. subjects and concepts, reconstructing their philosophical foundations, although, as Hider writes, the bibliographic current seems to be predominant anyway.

In a text from 2003, Hjørland focuses on defining the idea of a concept, but still doesn't abandon his initial approach, which appears to be stubbornly underlying his line of thought, embedded at the heart of its inspiration. This approach is the perception of knowledge as a factor that organizes the world. This organization, this ordering is understood literally: as a process of arranging things, establishing relationships, making various elements interdependent. These interdependencies are graspable—they are what knowledge organization is concerned with, and they exist in the world, although their existence is strongly linked to humans. Hjørland modifies the Platonic definition of knowledge from *Theaetetus* in that he introduces the idea of

[5] https://www.isko.org/index.php (accessed: 30.03.2021).

'knowledge claims' instead of 'knowledge' *per se* (Hjørland 2003, 100). This lets him justify the multiplicity of both the interpretations of the world within knowledge (many kinds of knowledge about the same subject, so to speak) and the variety of ordering structures (e.g. concepts) used by different sciences or approaches. This is also the answer to the question of why the world of knowledge is so rich, dense, and complex.

Knowledge is as if an ordering system itself. And there can be many of such systems. However, by changing Plato's definition, Hjørland deprives himself, a quite risky move, of the strong and well-thought-out basis for the existence of knowledge, which the philosopher saw to be in humans. Moving it towards claims and further—towards concepts—immediately raises the question about these concepts and about the way they exist. Knowledge organization is unable to free itself from this vicious circle, circulating obstinately at the level of said concepts. The assumption that knowledge is only claims locks this knowledge in a hell of illusions, because if there is something beyond them, some separate context of the network of relations, one that makes it possible to see this this network, it remains completely elusive for this knowledge.

It is indeed a frightening vision. A vision that must have resulted from the harrowing experience of a librarian, versed in the labyrinthine structure of knowledge like nobody else. After all, it exists, on the one hand, simply as a large collection of records, but on the other—this collection is humongous and grows uncontrollably and rapidly. Becoming a guide through this maze is a mission impossible to accomplish. The image of the knowledge that lies hidden there is the maze in question itself. But it is at the same time an act of surrender to knowledge and its nature.

The two types of instrumentalization of knowledge briefly presented, i.e. Knowledge Management (KM) and Knowledge Organization (KO), share a similar fate. Being promising tools for knowledge exploitation, they are subject to ongoing development and improvement. Both, however, experience a similar kind of defeat. A defeat that does not necessarily undermine their utility, because it is one that concerns a deeper level of their understanding. An understanding that is, after all, unnecessary from a utilitarian point of view. They are used because they offer a technical solution suitable for practical purposes. Both withdraw simultaneously from the opportunities that stand before them and act as doorways to huge intellectual gaps. The first, Knowledge Management, retreats from the cosmic dimension of meanings that make up the peculiar, exotic knowledge, which is a fantastic shortcut between the human and the universe. The other, Knowledge Organization, surrendered to knowledge, which appears as an abstract network of its own articulations, a multidimensional, complex, huge network. The former made do with pure practical exploitation, the latter focused on the practice of navigating the network of knowledge and, entangled, captured in this network, circulates through its connections and offers ever-new ideas for structures to organize this network. Both have stopped at a certain level of interpretation and, ultimately, only partially accomplished the task of capturing knowledge. So it looks like there's a need for a new tool to overcome this limitation.

References

Bell, Daniel. 1973. *The coming of post-industrial society: A venture in social forecasting*. 1st ed. New York: Basic Books.

Borowski, Andrzej. 2009. Humanizm Jako Przedmiot Współczesnej Humanistyki. In *Humanizm: Historie Pojęcia*, Humanizm. Syntezy, ed. Andrzej Borowski, vol. 2. Warszawa: Wydawnictwo Neriton.

Chandler, David, and Christian Fuchs, eds. 2019. *Digital objects, digital subjects: Interdisciplinary perspectives on capitalism, labour and politics in the age of big data*. London: University of Westminster Press.

Dalkir, Kimiz. 2005. *Knowledge management in theory and practice*. Boston: Elsevier/Butterworth Heinemann.

Davis, Joseph, Eswaran Subrahmanian, and Art Westerberg. 2005. Knowledge management: Conceptual foundations, emerging directions. In *Knowledge management: Organizational and technological dimensions*, ed. Joseph Davis, Eswaran Subrahmanian, and Art Westerberg, 3–20. Heidelberg: Physica-Verlag HD. https://doi.org/10.1007/3-7908-1618-3_1.

Hider, Philip. 2018. The terminological and disciplinary origins of information and knowledge organization. *Education for Information* 34 (2): 135–161. https://doi.org/10.3233/efi-180165.

Hislop, Donald, Peter A. Murray, Anup Shrestha, Jawad Syed, and Yusra Mouzughi. 2018. Knowledge management: (potential) Future research directions. In *The Palgrave handbook of knowledge management*, ed. Jawad Syed, Peter A. Murray, Donald Hislop, and Yusra Mouzughi, 691–703. Cham: Springer International Publishing. https://doi.org/10.1007/978-3-319-71434-9_28.

Hjørland, Birger. 2003. Fundamentals of knowledge organization. *Knowledge Organization* 30 (2): 87–111.

———. 2016. Knowledge organization. *Knowledge Organization* 43 (6): 475–484.

Kahrens, Marion, and Dieter H. Früauff. 2018. Critical evaluation of Nonaka's SECI model. In *The Palgrave handbook of knowledge management*, ed. Jawad Syed, Peter A. Murray, Donald Hislop, and Yusra Mouzughi, 53–83. Cham: Springer International Publishing. https://doi.org/10.1007/978-3-319-71434-9_3.

Kline, Morris. 1990. *Mathematical thought from ancient to modern times*. Vol. 3. New York: Oxford University Press.

Knorr-Cetina, Karin. 1981. *The manufacture of knowledge: An essay on the constructivist and contextual nature of science*. Elsevier Science Limited.

Latour, Bruno, and Steve Woolgar. 1979. *Laboratory life: The construction of scientific facts*. Princeton: Princeton University Press.

Lyotard, Jean-François. 1979. *La condition postmoderne. Rapport sur le savoir*. Paris: Éd. de minuit.

———. 1989. *The postmodern condition: A report on knowledge*. Manchester: University Press.

Maciąg, Rafał. 2012. *Deus ex machina: nowe media i ich projekt poznawczy*. Kraków: Wydawnictwo Uniwersytetu Jagiellońskiego.

Nonaka, Ikujiro. 1991. Knowledge-creating company. *Harvard Business Review* 69 (6): 96–104.

Nonaka, Ikujiro, and Hirotaka Takeuchi. 1991. *The knowledge-creating company how Japanese companies create the dynamics of innovation*. New York: Oxford University Press.

Nora, Simon, and Alain Minc. 1978. *L'informatisation de la Société*. Paris: La documentation française.

———. 1980a. *The computerization of society: A report to the president of France*. Cambridge, MA: MIT Press.

———. 1980b. Computerizing society. *Society* 17 (2): 25–30. https://doi.org/10.1007/BF02700056.

Polanyi, Michael. 1966. *The tacit dimension*. Chicago: University of Chicago Press.

Popper, Karl. 1935. *Logik der Forschung: Zur Erkenntnistheorie der Modernen Naturwissenschaft*. Vienna: Springer Vienna.

———. 2002. *The logic of scientific discovery: On the epistemology of modern science*. 2nd ed. London: Routledge.
Ryle Gilbert. 1949. *The concept of mind.*
Syed, Jawad, Peter A. Murray, Donald Hislop, and Yusra Mouzughi, eds. 2018. *The Palgrave handbook of knowledge management*. Cham: Springer International Publishing. https://doi.org/10.1007/978-3-319-71434-9.
Webster, Frank. 2006. *Theories of the information society*. 3rd ed. Abingdon: Routledge.
Wiig, Karl M. 1997. Knowledge management: Where did it come from and where will it go? *Expert Systems with Applications* 13 (1): 1–14. https://doi.org/10.1016/S0957-4174(97)00018-3.
Wittgenstein, Ludwig. 1958. *Philosophical investigations*. 2nd ed. Oxford: Basil Blackwell.

Chapter 9
Boole and the Universe of Thought

The five stories presented so far, featuring—in order—Gettier, Fleck, Masuda, Foucault, and Lyotard, point to a kind of dualism in the approach to knowledge. On the one hand, knowledge is treated as a certain whole, as a transcendent space that can be identified with the world as such, being a representation of this world. The nature of this knowledge is vague. But it certainly has a mysterious, deep bond with the human man, who is considered its 'possessor' and 'owner' for this very reason. This image becomes gradually dispersed and, viewed in a variety of contexts, loses its basic characteristics, offering a picture of knowledge that is extremely different. Such knowledge thus presents itself primarily as a local event, fixed in certain peculiar circumstances, among which the most obvious are social circumstances. The latter must become important because knowledge turns out to be determined by the act of communication, which gradually embraces its specificity in full. It is further dispersed in the sense that it emerges as a set, and in the sense that it is a set of parts, a collection of elements being its various manifestations that overlap or intersect with each other. Knowledge exists then as a mixture—a chaotic one, because it depends on diverse and numerous social determinants.

Thus, locality affects the cohesiveness of the matter that forms knowledge, which reveals itself in simplest terms as its temporality. Knowledge remains captive to time—which can also be a historical timeline, i.e. determined by its cultural, social, and all other properties. This way, knowledge gets rid of its nature deemed to transcend beyond humans and their world. This process, paradoxically, results in the removal of the human being as the main individual subject of knowledge. The human being is substituted with a collective subject, which is also developing in an interesting way nowadays, in an online version. This takes place in line with the metaphorical and universal vision offered by Foucault, in which the individualistic concept of man as the foundation of history turned out to be merely a rhetorical figure, an intellectual construction, a footprint in the sand. This transformation, however, is much more common and widespread, and is emerging as one of the dominant themes of the social sciences in the second half of the twentieth century.

R. Maciag, *Knowledge as a Tale*, https://doi.org/10.1007/978-3-031-69820-0_9

A collective subject is a relatively new invention. If we were to measure its history by the science dealing with this aspect—sociology, it can be said that it was born as a self-conscious being in the nineteenth century through August Comte and Herbert Spencer (Maus 1962; Swingewood 1991). It becomes 'graspable' in absolute terms thanks to a tool that emerges at the beginning of the twentieth century—the concept of networks, which introduces the possibility of describing any human collective of a strict nature, based on Euler's mathematical concept of graph. This tool develops and reaches maturity at the very end of the twentieth century, which results in a number of its applications in various fields of science, producing the impression that the tool is universal since the nature of the world seems intrinsically networked. Among said applications, the description and modeling of human collectivities is flourishing on its own, gaining a strong boost from the vast amount of data of this kind, collected through social networks built with the help of the Internet. The latter phenomenon, thanks to the extraordinary dynamics of its development and to its enormous scale, became the most important subject of analysis into human collectivities around the year 2010. At the same time, the idea of collectivity, hitherto difficult to grasp research-wise, empirically, materializes directly as a growing body of data representing this collectivity in various increasingly refined ways. Thanks to the fact that they are saved, recorded, and collected, they become a whole new reservoir offering access to the world of human relations hitherto unseen, sensitive, intimate, and so on. This process makes the collective seem more visible and accessible today than ever (Golbeck 2013).

Knowledge turns out to be more and more an interpersonal experience, disrupting its previously clear, unambiguous status of man's relationship with the world. The latter relationship is not discarded, of course, but the filth of social relations makes its way into its pristine homogeneity. The aforementioned homogeneity was exercised simultaneously both as an affirmation of man and as an affirmation of the world, though not necessarily at the same time, spawning sublime philosophical forms of description and generating virtually all known philosophy. Adding a drop of foreign color ingredient to this pure solution gives rise to unexpected effects. This interference is practiced in various ways—described here already—but the result is always the same: there is no way to keep knowledge pure. Knowledge loses its untouchability and superiority at the same time. It mixes irreversibly with the harsh dirt of everyday life.

This dirt that comes with the reality of social relationships is, of course, inequality and domination inherent in them. The exploitation of knowledge within the framework of so-called Knowledge Management (KM) is perhaps the most visible and obvious form of gaining a better and higher position in the race called competition. The relationship between knowledge and power, one discovered to exist already by Plato, takes on an entirely new nature. It is precisely such a relationship that is the subject of the close analysis proposed by Foucault. In this case, power is strikingly trivial. A much more complex image of these relationships is offered by Magala (2021). In light of the latter analysis, the gradual socialization of knowledge and the loss of its idealistic universality, as recounted here, results in an idyllic and innocent picture. But today's forms of exploitation of knowledge as a social resource have expanded to an extremely advanced—even violent—degree.

The problem of communication, left aside, has its own remarkable back story, which we discover thanks to Michel Foucault. At a similar time when the way people are viewed as part of a collective (which requires a separate science—sociology) transforms, there is also a profound change in attitudes toward language. Foucault describes this process in *The Order of Things*. This change is very similar to the one discussed above. It takes effect as the objectification of language and deprivation thereof of its hitherto transparent nature, placing it above the world it represents. This unique and superior status determines its special role in the process of cognition and breaks down in the nineteenth century. Language becomes one of the many things that are subjected to careful and disciplined examination—as are the interpersonal relationships that produce various collectivities and give them a variety of dynamics. Foucault writes: "In the seventeenth and eighteenth centuries, it [language – author's note] was the immediate and spontaneous unfolding of representations; it was in that order in the first place that representations received their primary signs, patterned and re-grouped their common features, and established their relations of identity or attribution; language was a form of knowing and knowing was automatically discourse. Thus, language occupied a fundamental situation in relation to all knowledge: it was only by the medium of language that the things of the world could be known. [...] From the nineteenth century, language began to fold in upon itself, to acquire its own particular density, to deploy a history, an objectivity, and laws of its own. It became one object of knowledge among others, on the same level as living beings, wealth and value, and the history of events and men. It may possess its own concepts, but the analyses that bear upon it have their roots at the same level as those that deal with other empirical forms of knowledge. The pre-eminence that enabled general grammar to be logic while at the same time intersecting with it has now been lost. To know language is no longer to come as close as possible to knowledge itself; it is merely to apply the methods of understanding in general to a particular domain of objectivity" (Foucault 2002, 322).

Language is crucial here, as it is designated as the primary cognitive tool, which is first treated as granted, transparent, obvious—and thus not requiring attention. It is like air: necessary to live, but invisible. In this approach, which is obviously gullible, we have to quickly reconsider and subject language to a separate examination. According to Foucault, this is what happens in the nineteenth century. It's hard to resist the perspicacity of his analysis of the process within which words and things, directly connected with each other, were arranged. In the seventeenth and eighteenth centuries this relationship was considered natural and necessary. Things made themselves known through language, which acted as their natural surface facing man. It was therefore, as Foucault writes, a pure form of knowing. This surface appeared to be orderly, organized, and reasonable, which meant a certain progress compared to the Renaissance. The nineteenth century ushered in and embraced a change, a continuation of the historical development of the way of thinking about the role of language. Language, as Foucault argues in his remarkable synthesis, belonged to the world, it was hidden inside its entities. As the story continued, it gradually came to the foreground. In the nineteenth century, it reached the status of a separate subject, without losing its cognitive function. It thus became a tool, one subjected to a study that analyzed its practicality and utility.

Foucault sees this process as a deprecation of language, relegated among other objects, and refers to its unique properties that let it play a cognitive—unique—role as forms of compensation. This somewhat poetic treatment stems from the aforementioned uniqueness of language's position among other things—after all, as a cognitive tool, it faces these things (and itself). There are three ways in which the demotion of language to the status of an object is compensated for: firstly, as Foucault claims, "by the fact that it is a necessary medium for any scientific knowledge that wishes to be expressed in discourse" (Foucault 2002, 303). Secondly, language becomes a study subject because it is an extraordinary resource on account of its unique position. Exploration of this resource, philology, hermeneutics, and interpretation produce powerful areas of reflection. Thirdly—and surprisingly, the act of reclaiming language through the pragmatic procedures of its use results in the uncovering of its extraordinary power and profligacy of its basic components: words. This condensed energy leads to an explosion of the pure matter of language—literature.

From the point of view of this way of thinking, the first form of compensation is the most important—not only because it regulates the processes of cognition, i.e. the emergence of knowledge, which now must involve language, since it is language that proves necessary for these processes to be carried out. This is also because the result of this activity, which language must perform, is two more consequences. The first, quite simple and obvious, is about an attempt to adopt language to the role of a tool of cognition, which is to make it as clear and as uninfluenced by the represented object as possible. The second is actually a continuation of the first. It is a search for the pre-linguistic order that language relies on when it makes use of it, imposing its own rules of the game, i.e. grammar, vocabularies, etc. Foucault points to such an order and gives an example thereof. It is logic, and its most prominent advocate is George Boole, the inventor of the calculus of logic. Foucault interprets this case as a response to the need to construct a pure symbolicity, a formalization that acts as a safe and organized foundation for the unbridled, incalculable, and bright matter of language.

The concept of how language works proposed by Foucault now leads us down two paths. First, it is consistent in building a certain interpretation of this phenomenon, taking on a role that is absolutely fundamental to the process of recognizing and arranging the dimension of things. It represents one of the most important points in the development of this thought in the twentieth century, featuring Wittgenstein at its beginning, but also a number of other great scholars of language, such as de Saussure, continued later on by Austin, Rorty, and Derrida.[1] The central

[1] The inclusion of analytic philosophy into this tradition shifts the origins of the modern discussion of language and its cognitive role to Frege (Dummett 1994) and generally paves the way to the problem of the so-called "linguistic turn", commonly known by this name thanks to Rorty's book (Rorty 1992 [1967]).

figures to the story of this development must make an open bet in a certain betting game, which Badiou succinctly described as follows: "either truth or the arbitrariness of the sign and the diversity of syntactical games: this is the central choice for contemporary philosophy" (Badiou 2008, 49). Foucault and Lyotard choose the latter, at least in the part of their output cited here. And, in consequence, reject the truth. Second, by invoking the figure of Boole, Foucault points us to a relatively new phenomenon that will have a primary application here—the phenomenon of discourse. The relevance of Boole to the construction of discourse is fundamental and needs to be addressed in more detail.

Badiou and Language as an Analogy of Mathematics

First in line is, however, Badiou's thought, which has been somewhat taken out of context, because it appears as part of his musings on number, in a part where he deals with Peano's reflections. Such a context proposes a slightly different approach to the understanding of the idea behind the presented view of how language functions as a cognitive tool. The idea in question omits the entire multifaceted discussion accompanying the aforementioned "linguistic turn", but sees the similarity that occurs here. Even more than a similarity, in fact: a kind of morphological affinity between mathematical structures and language in the versions advocated for by Peano and Lyotard, respectively, with which, incidentally, it disagrees. The value of this reasoning for the argument of this book, however, lies in the possibility of thinking of language as an equivalent of a mathematical notation. Badiou examines this juxtaposition mainly from the perspective of mathematics, which he deals with. Yet, the other aspect remains. The aspect of language, which, according to his argument, can, under certain conditions, be juxtaposed with mathematics and make it possible to come up with analogies. Badiou thus shows—somewhat contradicting himself—a potential avenue for interpreting the descriptive (though, after all, also cognitive) role of language, without having to get tangled in complex analyses of its functioning. Without linguistics, philology or hermeneutics, but by recognizing its potential power as a formalized syntactic system. And that's completely enough for us.

The deep source of all reasoning is the crisis of the philosophy of mathematics, or—more precisely,—of its ontological foundations. This crisis materializes itself as fundamental questions about how the basic entities of mathematics, including—above all—numbers, exist. Among the philosophers answering these questions, the following three are mentioned most often in this context: Gottlob Frege, Richard Dedekind, and Giuseppe Peano, with the latter considered the most advanced innovator, whose idea became the basis for the reconstruction of thinking about mathematics in the twentieth century.[2] Its meaning boils down precisely to the understanding of number, but also all —mathematical sentences as formal constructs, defined according to circumstances by a system of assumptions (axioms) and devoid of any external justifications. Badiou summarizes this attitude as follows: "Peano does not broach questions of existence. When a system of axioms is applied to operational arrangements, we will be able, if necessary, to enquire as to that system's coherence; we need not speculate on the being of that which is interrogated. [...] He thus participates forcefully in that movement of thought, victorious today, that wrests mathematics from its antique philosophical pedestal and represents it to us as a grammar of signs where all that matters is the making explicit of the code" (Badiou 2008, 48).

It is this kind of thinking to which Badiou compares the "linguistic turn", a phenomenon he explains through the meaning Lyotard gives to it. Lyotard, the figure covered in the previous chapter, uses the term very rarely. Once, however, he discusses it in a properly comprehensive—albeit brief—manner. This happens in a kind of foreword to one of his most important books: *Le Differénd* (Lyotard 1983). This work starts with a chapter titled *Reading Dossier*, which is a kind of summary of the argument offered in the book, consisting of a number of organizational matters, among which there is a separate section titled *Context*, devoted completely to this very phrase. Let me quote it almost in its entirety: "The "linguistic turn" of Western philosophy (Heidegger's later works, the penetration of Anglo-American philosophies into European thought, the development of language technologies); and correlatively, the decline of universalistic discourses (the metaphysical doctrines of modern times: narratives of progress, of socialism, of abundance, of knowledge). The weariness with regard to "theory" and the miserable slackening that goes along with it (new this, new that, post-this, post-that, etc.) (Lyotard 1988, viii).

In addition to the aforementioned three direct sources, the definition offers a generalization thereof, which at the same time acts as a direct crux of the matter (on account of the magic word "correlatively"). It is this "decline of metaphysical discourses" that is also the most important theme of *The Postmodern Condition*, which we have dealt with in the previous chapter. It seems that this is the thought that Badiou refers to, as it indeed shows a similar approach. The formalized axiomatic systems, which are basically arbitrary but extremely disciplined products of imagination, as referred to by Badiou, are analogous to these discourses—which are merely free constructions, albeit perceived erroneously so far as different emanations of forces external to man. They are described more extensively in *The Postmodern Condition*. If they are indeed casual and temporary, they cannot aspire to be explanatory "theories"—inherently selfish, haughty, and intolerant of competition. Thus, they turn out to be rather makeshift descriptive constructs, which means they offer only a certain way (system) of description, allowing the use of syntax and grammar chosen on an *ad hoc* basis.

[2] This is a constantly recurring context, which has already appeared here in Chaps. 5, 6, and 7—and will appear again. Not surprisingly, it is the most important source of the cases of knowledge discussed here.

The similarity of the nature of language to mathematics as it emerged at the end of the nineteenth century is underlined quite frequently across Lyotard's work, and is generally accompanied by a reference to the technologies of the time called "information technologies" as a direct reference to Shannon and Weaver's idea. Badiou speaks of this similarity directly: "We see here, as if in the pangs of its birth, the real origin of what Lyotard calls the 'linguistic turn' in Western philosophy, and what I call the reign of the great modern sophistry: if it is true that mathematics, the highest expression of pure thought, in the final analysis consists of nothing but syntactical apparatuses, grammars of signs, then a fortiori all thought falls under the constitutive rule of language" (Badiou 2008, 48). This observation refers to a clear alternative that is explicitly constructed, explaining the content of the wager mentioned at the beginning of this passage: either truth, the Platonic belief in the union of things and words, and the ontological hardness of number, or "arbitrariness of the sign and the diversity of syntactical games", words as free signs and number as an ephemeral element of mathematical syntactics. We don't need to choose. The decision has already been made. The transformation of language, which Foucault describes, is about language losing its idealistic foundation, its strict logic of representation. It then becomes ultimately viewed as a changeable object of constant observation, manifesting itself differently each time and according to local circumstances. Badiou acknowledges that in order to understand this process, it is enough to rely on an analogous and much more fundamental transformation of the perception of mathematics, which turns from insight into the essence of things into sentences and the rules of formulating them. In this situation, the concept of truth becomes irrelevant. Kline calls this process, which becomes a historical fact, a "loss of truth" (Kline 1990), making a choice in the process that Badiou described as "central choice for contemporary philosophy". The choice made by Lyotard, Foucault, and others was thus a logical consequence of the process that took place in the domain of mathematics. Its history, on the other hand, provided the right amount of justification and explanation for it. Badiou comes forward with further arguments for this.

The first path we follow guided by Foucault leads us to a reasoned acceptance of the notion of language as a set of temporary, ad hoc, local rules that control its order, adapted accordingly to the task this language faces. When it is to serve as a tool to describe the world in scientific terms, it seeks, as Foucault argues, the most compliant and appropriate solutions, or it tries the mathematical way: it appeals to formalized logic as a completely transparent and abstract order. In presenting this very sophisticated and general synthesis, Foucault gives the example of Boole's calculus of logic in the latter case, and this makes it simultaneously possible to proceed naturally and consistently with the necessary description and offer a plausible justification of the construction of fundamental importance here. For discourse, which is the idea in question here, was born as a disciplined form more than a century before it became 'elevated' and glamorized by Foucault. And it was a birth that heralded a change in the world. A change in how we think about it.

Boole and the Decomposition of the World

George Boole is a remarkable figure from the point of view of the history of mathematics, or, speaking more precisely, of mathematical logic because his ideas have in fact shaped today's world viewed as a result of the invention of the computer. After all, the operation of this device is based on his logic. From the point of view of this book, however, his role and his impact on knowledge is different. He lays the foundation for looking at the world in a way that is useful for building a certain model of knowledge, which I refer to as discursive knowledge.

Rolf George and James van Evra describe Boole's contribution as an idea based on the subordination of logic to mathematics instead of philosophy. This was based on an idea discontinued in 1847 in Boole's book *Mathematical Analysis of Logic*, where he introduced the concept of an "*elective symbol*", for instance an *x*, which, as George and van Evra write, "represents the result of 'electing' x from the universe" (George and van Evra 2006, 41). This way, a class of objects represented by said x comes to being. The idea of using letters to denote such classes came from Boole, who used them in the same way they had appeared by then in algebra and started doing calculations on them. However, the decision to label any entities, including the world of objects, using mathematical methods had surprising consequences, for it led to the conclusion that the algebra created in this way must be limited to only two values—0 and 1, which follows directly from the value that existence can take traditionally, that is, according to Aristotle's reasoning.[3] Thus, Boolean logic is born as an idea of formalizing basic ontological qualifications, and thus already contains an immanent ambition to embrace the real world, which was a grand, significant act, one that actually opened the way to the question of the interrelationship of the digital world and the real world, encompassing both meanings of 'digitality': narrow, concerning computer technology, and broad—meaning a cognitive paradigm based on an advanced interpretation of mathematics and its non-simple relationship with the world as the object it describes. This act is also the first step towards sorting out the relationship of mathematics to the real world—a problem that is to about to turn into a violent crisis at the end of the nineteenth century.

The evidence of the importance Boole attached to his concept can be seen clearly in the first words of his second book, *An Investigation of the Laws of Thought* (Boole 1854). It opens with the following statement: "The design of the following treatise is to investigate the fundamental laws of those operations of the mind by which reasoning is performed" (Boole 1854, 1). His achievement, as he claims, will have several intended effects, with the first being to express these laws in the form of a Calculus and base a certain Logic on it (in Boole's original text, the words 'calculus' and 'logic' are capitalized). This goal, referring directly to the dream of Leibniz and Descartes,[4] is consistently maintained, producing a constant tension between the reality of formal calculus and human reasoning, which is represented by thought.

---

[3] Davis discusses this issue in an accessible and extensive manner (Davis 2000)

[4] I address these issues in (Maciąg 2012).

At the same time, Boole keeps his course steady, departing in his search from reasoning. Its certain practicality of consistently relying on the world of phenomena yet to be rendered in the abstraction of logical calculus leads Boole to understand language in strict terms as "*an instrument of reasoning*". This type of language, according to the aforementioned tension, is accompanied by its colloquial source version. It is, of course, necessary to draw a fine line between these two linguistic realities if the goal is to codify the rules governing the use of formal language, the beginnings of which we have just come to see. This line is drawn by means of limiting possibilities and entities, and standardizing notions, which become terms. This necessity, however, has the effect of interfering with the world of thought, to use Boole's language. After all, there is no chance that such concepts will retain their previous wealth of associations and meanings. By design, they become a certain choice. A reduction that produces a certain hypothetical-only whole is necessary (Nambiar 2000). It is this necessity that underlies the introduction of the construct the idea of which Boole borrowed from his slightly older friend Augustus De Morgan. Boole refers to this construct as "the universe of discourse". De Morgan introduced the concept of universe in his 1846 text *On the Syllogism: I*. This universe was to be a kind of arbitrary and assumed set of entities chosen for reasoning (De Morgan 1966, 2).

Boole elaborates on and clarifies De Morgan's rather casual observation, following a line of thought that is simple and reasonable in light of his endeavors: In every discourse, whether of the mind conversing with its own thoughts or of the individual in his intercourse with others, there is an assumed or expressed limit within which the subjects of its operation are confined. The most unfettered discourse is that in which the words we use are understood in the widest possible application, and for them the limits of discourse are co-extensive with those of the universe itself. But more usually we confine ourselves to a less spacious field. Sometimes, in the discoursing of men we imply (without expressing the limitation) that it is of men only under certain circumstances and conditions that we speak, as of civilized men, or of men in the vigor of life, or of men under some other condition or relation. Now, whatever may be the extent of the field within which all the objects of our discourse are found, that field may properly be termed the universe of discourse" (Boole 1854, 42).

This extensive observation and the conclusion that follows hold a very basic idea that will reveal its potential in a moment. Its strength lies in its ability to conceptualize the real world according to the conditions imposed by an external local order, which implies an approach opposite to the current one. It is no longer the world that sets the standards for cognition. It is the procedures of that cognition that require assumptions about the world, which must ultimately lead to the world's deformation, so to speak. The world becomes an object of manipulation and loses the hardness of the untouchable being it has seemed to be. In particular, such manipulation can involve any description order (e.g. logic), justified only by the procedure itself. Such ordering is, of course, a type of knowledge, as clearly exemplified by the field of research called Knowledge Organization, which I discussed in the previous chapter.

Boole also naturally identifies the world, which is the subject of reduction, with the story of it, i.e. with language. And he does it many years before the mature form of this identification is proposed by Wittgenstein, discussing with himself, in the next century. In specifying the conditions under which the phenomenon of arbitrarily limited discourse appears, he writes as follows: "The office of any name or descriptive term employed under the limitations supposed is not to raise in the mind the conception of all the beings or objects to which that name or description is applicable, but only of those which exist within the supposed universe of discourse" (Boole 1854, 42). The last quote is among Boole's oft-quoted words. With this interpretation, necessary from the point of view of Boolean reasoning, an entity emerges on which many schools of his interpretation will rely: discourse as a closed, confined space, which can be, of course, one of many similar spaces. The notion of space is used here in an exaggerated way and will be specified in the next chapter. Its purpose is to highlight the emancipation of discourse, which takes place through its potentially numerous instances—which are, by definition, limited and separate. This results in a picture of the possible multiplicity and coexistence of a divided discourse or—perhaps—a number of discourses.

Boole's use of the word 'universe', borrowed from De Morgan, is rather paradoxically associated with boundlessness, which is probably a legacy of the earlier image of knowledge as a generalized interpretation of the world. At the time, knowledge seemed to be a homogeneous, unified discipline. However, the universe of discourse produces a completely different, discrete, granular, diffuse image. This is due to the practical arbitrariness of determining the domain of the universe of discourse. Keynes emphasizes this, arguing that "no general criterion can be laid down for determining what is actually the universe of discourse in any particular case. It may, however, be said that knowledge as to what is the universe referred to is involved in understanding the meaning of any given proposition; and cases in which there can be any practical doubt are exceptional" (Keynes 1900, 213). Allowing for the possibility of an infinite dispersion of parallel universes of discourse ultimately also suggests the possibility of existence of different instances of knowledge.

But it needs to be added, however, that the universe of discourse, in its potential boundlessness, does not confine itself to any entities, which causes some difficulty for Keynes when he tries to analyze the situation in which creations of fiction, such as literary characters, are to take on the role of elements of the universe of discourse; creations whose ontological status would be analyzed only many years later by Roman Ingarden, as discussed to in Chap. 1. In light of the above, Keynes sticks to the assumption that the basis of such a universe must be "*real content*" (Keynes 1900, 211). Boole is not that overbearing, though. Considering the seemingly transparent layer of language—or, more precisely, that part of it which, following de Saussure, we would call *signifié—the signifier*, the concept, emphasizes the participation of mental operations in the creation of the names used in logical sentences. He sees it very clearly and highlights the role of the "*powers and limitations of the human mind*" in the production of certain *conceptions* of these names, and there is no doubt that we are dealing here with an intellectual act (Boole 1854, 43). Boole thus takes the problem of the modality of denotative names into the realm of entities of a completely different kind, inhabiting the "world of thought", which is, after all, the main subject of his deliberations.

Modern definitions of the universe of discourse do not address this issue, doing with very general properties and not offering any preliminary ontological restrictions. This seems to be a consequence of the great discussion that took place in the twentieth century. A discussion that considered the various interpretations of the status of existence of different entities, with one of these interpretation being Ingarden's aforementioned idea. The Cambridge Dictionary of Philosophy offers the following definition: "The universe of discourse of a given discussion, treatise, or discourse, is the usually-limited class of individuals under discussion, whose existence is presupposed by the discussants, and which in some sense constitute the ultimate subject-matter of the discussion. Once the universe of a discourse has been established, expressions such as 'every object' and 'some object' refer respectively to every object in the universe of discourse and to some object in the universe of discourse" (Audi 1999, 941). George and van Evra, in their book *A Companion to Philosophical Logic*, define it in passing, treating it as a well-known, common tool that needs no broader description. According to them, the universe of discourse is "still generally used, as a way of targeting statements to a class of objects under discussion, rather than the entire universe" (George and van Evra 2006, 42). It is defined slightly differently in the Computer Science and Communications Dictionary. Apart from the usual characteristics, presented in a remarkably succinct manner, the definition includes also additional capabilities based on other procedures for determining sets of beings: "in a particular context, all of the entities or elements that (a) are of interest and (b) may include (i) many entity worlds and (ii) entities that are not yet perceived or considered" (Weik 2001). However, all of these definitions do not refer in any way to the individuals, objects or entities of which they speak—even though they name them in different ways. Neither do they determine their ontological foundations or simply the form of their existence.

Boole's approach was much more realistic, which may seem a bit gullible today. It certainly comes from unfamiliarity of the advances made after him in the field of language, discourse or ontology. However, the fields of reflection he opened up have grown tremendously, creating their own fields of research. Indeed, discourse became an extremely broad area of research and application in the twentieth century. But before we deal with it in more detail, let's keep in mind its hidden foundation, which appeared in Boole's brilliant proposition. Choice. The universe of discourse is the result of each choice made from the multitude of options the world offers. Constructed to achieve a certain pragmatic purpose. It is completely irrelevant whether this purpose is solely the result of a deliberate decision or an accepted existing state. The universe as a construct is thus a creation with no claim to generality. No claim to exclusivity. No claim to singularity. Instead, it is an open step to a diffuse, granular, discontinuous reality, but also a reality that is disordered in the sense that the separated parts don't have to be mutually exclusive at all. On the contrary. They can overlap in an infinite number of combinations. This is the reality of multiplicity of arbitrary universes. Looking from this point of view, the universe of discourse is an image of the world yet to come, and the source of the question about the nature—the method—of formation its parts.

And the second thing. The creation and isolation of separate and arbitrary wholes in Boole's construction takes place as an interpretive, theoretical, speculative, thinking experience, because such a decision lies at its core. This way, Boole introduces us to the twentieth century and its cognitive structures. The very same assumptions make it possible to build Peano's and Hilbert's axiomatic systems—this is the end of the nineteenth century. Another extension of this idea bears fruits in the form of the concepts of Mach and Einstein. We move freely between the fields of science, mathematics, and physics. But this is legitimate because in each of these concepts, one tension is constant, though understood differently. It is the tension between thought and the real world, which is the object of that thought.

For Peano and Hilbert, the real world is not necessary at all. They are happy with purely speculative and abstract constructs. They produce their own worlds, the legitimacy of which comes solely from within themselves, as a materialization of the assumed axioms that brought these "worlds" to life. Mach and Einstein deal with the physical world, which ultimately turns out to be governed by a similar principle. They are not moving in the safe zone of speculation. Their discovery concerns the experiential world and its fundamental, basic order. The order of the world is described in physics using entities such as space and time. Mach and Einstein discover that space and time do not exist absolutely and independently, but relativistically. Cheng writes: "Space and time should not be like a stage upon which physical events take place, and thus have an existence even in the absence of physical interactions. In Mach's and Einstein's view, space and time are nothing but expressing relationships among physical processes in the world—"space is not a thing."" (Cheng 2005, 10).

To Newton, the world seemed a universal and eternal environment of things and processes. Descartes referred to it as *res extensa*—a continuous and homogeneous thing, standing in opposition to *res cogitans*, a second territory, at the exclusive disposal of man. The ability that makes this juxtaposition possible is a skill exclusive and unique to humans: thinking. Both were convinced that there is only one physical world in the sense that everything exists according to a single order that does not depend on anything. It is first, main, prior, and indivisible. The relativist approach makes such an outlook irrelevant and lends world phenomena the characteristic of relativity, which means a multiplicity of possible coordinate systems—meaning a multiplicity of relative interpretations. The coherence of such a world is preserved by the fact that changing the coordinate system does not affect the behavior of its components; it is only the context of their description that changes.

The essential premise of these unusual constructions, i.e. the need to adopt initial and later valid guidelines for description, appears precisely in Boole's writing. And it is a matter of choice made by the author of these constructions, who can impose different degrees of freedom on this choice depending on the purpose. In the world of speculation, such an author is the creator of these constructions. In the physical world, the author is an observer thereof. Boole's idea grows into a third powerful area of constructions that position themselves as if between pure speculation and experience, with their author being both the creator and the observer of discursive constructions. This time they appear to be certain specific hybrids: partly contrived and partly self-contained, autonomous. The field where this kind of duality can occur is language, a creation that remains in communication with the real world on the one hand, and with free thought, largely sovereign, on the other. Of course, no part of this duality is simple. Rather, these parts actually pave the way to further questions. And this is how the theory of discourse is produced to guide us towards knowledge.

The Development of the Tool of Discourse

Discourse has proven to be an extremely useful and handy tool for exploring the space of linguistic creations and—through these creations—the real world, as it is. The size and complexity of this field is addressed in numerous textbooks (De Fina and Georgakopoulou 2020; Gee and Handford 2013; Schiffrin et al. 2001; Schober et al. 2018) and general studies (Angermuller et al. 2014; Hyland and Paltridge 2011; Jørgensen and Phillips 2002). The name "*discourse analysis*" encompasses areas of science directly associated with language, such as linguistics, pragmatics, ethnography or media and communication. Yet, it reaches much further—into the realm of anthropology, history, psychology, literary studies, sociology or philosophy.

In his textbook, Paltridge describes discourse analysis as a research method that relies on language, but only to set out to explore its many entanglements in a variety of processes, which immediately makes it a very universal pretext to enter and observe the world. He writes that "it looks at patterns of language across texts and considers the relationship between language and the contexts in which it is used. Discourse analysis also considers the ways that the use of language presents different views of the world and different understandings. Thus, discourse analysis examines how the use of language is influenced by relationships between language users as well as the mutual effects the use of language has upon social identities and relations. It also considers how views of the world, identities, are constructed through the use of discourse" (Paltridge 2006, 2). It is clear from this description that it is not so much the language as the reality in which this language appears that is the object of study. Apart from the obvious social circumstances, it also encompasses a far deeper sphere—knowledge, meaning the perception of the world and its understanding. Viewing discourse this way must mean that the field of observation it opens up is exploding rapidly, embracing almost the entirety of the social sciences and a chunk of the humanities.

When Paltrigde goes back to the early days of discourse analysis, he discovers an interesting parallel. He mentions Zellig Harris as the originator of the term—and one to establish the basic principles of discourse in 1952 (Harris 1952). Harris is a formalist in his approach: he is supposed to be concerned with analyzing the occurrences of morphemes as selected linguistic particles and nothing more, but he still cannot free himself from the obvious presence of a context other than the purely linguistic, textual one: the imposing social circumstances. And thus he establishes, somewhat against his own pursuits, a field of discourse that goes far beyond descriptive linguistics. This is, of course, immediately picked up by Paltridge, who emphasizes two different ways of approaching discourse: first, formally, as an analysis of language in structures more complex than the sentence, and second, as an analysis of language in an extra-linguistic reality.

Harris is therefore the patron and originator of both paths that discourse analysis has taken. The formal path leads to the areas of speech and language processing, natural language processing, computational linguistics, and speech recognition. These elements are featured in the very well-known, extremely knowledge-rich—and still developing—textbook by Daniel Jurafsky and James Martin,[5] who mention that Harris was the author of the first machine syntax analyzer of sentences (parser). The device was created in the late 1950s as part of a project called the Transformations and Discourse Analysis Project (TDAP), and belongs today to the tradition of the field of research and technology as indicated by Jurafsky and Martin (Jurafsky and Martin 2009, 170). This field defines discourse analysis very precisely and in line with the tradition initiated by Harris: "Discourse analysis is the study of the structure, relations, and meaning in units of text that are longer than a single sentence. More specifically, it investigates the flow of information and meaning by a collection of sentences taken as a whole" (Kamath et al. 2019, 105). Discourse is defined very succinctly in this context. It's simply "structured, coherent groups of sentences" (Jurafsky and Martin 2009, 442).

Kamath et al. and Jurafsky and Martin treat discourse analysis as part of a broader problem that can be temporarily labeled as Natural Language Processing (NLP). Discourse, however, as an object of machine analysis, can appear as if on its own, precisely as a tool for social research. This approach has its own tradition, with its origins in the writings of Michel Pêcheux (Wiedemann 2016, 45) and his 1969 book *Analyse automatique du discours* (Automatic Discourse Analysis) (Pêcheux 1969). The two perspectives in approaching the use of computational techniques in the study of discourse, both the instrumental one, which uses discourse but with a more general task to handle, and the one that directly investigates discourse as a phenomenon, eventually meet. The latter even makes use of the former's connotations (Dascălu 2014). Jurafsky and Martin extensively discuss the historical processes of the development of the two, quoting a number of sources. An important thing to notice, however, is that with the development of their book, which is concerned with general problems of speech and language processing, in the latest, third edition thereof, they decide not to present discourse in the context of their own argument even generally. They only point to its features that they find important, especially coherence. This technical and instrumental approach is presumably a sign of the distinctiveness of their approach to the discourse analysis existing as a separate field of social research. The literature they cite changes diametrically too. Still, the

[5] A working version of the third edition is available at https://web.stanford.edu/~jurafsky/slp3/ (accessed: 26.04.2021).

tradition and contemporaneity of formal discourse analysis do not become any less intrinsically entangled because of this.

Let us recall that a trip into the complex and thick maze of discourse analysis subjected to machine (computational) tools includes only one way of analyzing it—although it is useful in studying the latter area as well. The very one that makes it the main object of its fascination and pursues it in all possible incarnations and relationships as a universal and synthetic insight into the world. Its most important—eventually defined—research territory is the so-called Discourse Studies. The originator and central figure of this discipline is Teun van Dijk, who in his inaugural 1997 collection describes it as a study that runs through all disciplines of the social sciences and humanities (Dijk 1997a). His introduction also offers a difficult (and long) answer to the question of what discourse is. Yet, this answer cannot be squeezed into just a few sentences. Discourse is inherently *fuzzy*, although it is possible to formulate three basic dimensions of it: a) it is a use of language, b) it is a way of communicating beliefs (cognition), and c) it is an interaction in social relationships (Dijk 1997b, 2). All three extend to form vast areas, encompassing the territories of many scientific fields.

Teun van Dijk points to the origins of discourse studies, which emerged in the mid-1960s. Angermuller et al., writing an introduction to one of the many later textbooks, trace its origins to the tradition of discourse analysis—a field that they believe to encompass all avenues of inquiry and an extremely heterogeneous field of research. By covering this field briefly, they portray it as a triad consisting of three dimensions, clearly based on van Dijk's proposal. These three dimensions are language, practice, and context (Angermuller et al. 2014, 7). However, this level of generality is justified by the presentation of the actual panorama of variety and riches hiding behind it. Angermuller et al. organize it based on five distinct criteria (Angermuller et al. 2014, 1). This heterogeneity is due, first, to the diversity of scientific and intellectual traditions—i.e. to the abundance of theoretical foundations of investigation, such as post-structuralism, semiotics, praxeology, hermeneutics, and many others upon which it can be justified. Second, its source is the dispersion of research disciplines within the boundaries of which scientific effort can take place. Third, this heterogeneity is the result of a number of interpretive schools, which came to being through the personal contributions of their prominent founders. Among these founders, Angermuller et al. list—in addition to Michel Foucault—the already well-known Michel Pêcheux, Teun van Dijk, Zellig Harris, as well as many, many other very different scholars. The fourth reason stems directly from the different types of data used as the basis for research. This data can be of a variety of types, extending the concept of text far beyond written sources. Reason five is also technical, though more general, and originates from a reality rich in both theoretical concepts about discourse and in empirical research concerned with observation of the world, supplementing speculative or philosophical foundations.

The above list is well summarized by the definition of discourse analysis as provided by Crane: "the theory and methodology for the study of the dynamics of human linguistic interaction and accomplishment" (Crane 2016, 89), or by Paltridge, who writes that in addition to language, the knowledge it provides simply encompasses the entire "world beyond word, clause, phrase and sentence" (Paltridge 2006, 20). Or by Gee and Hanford: "the study of language above the level of a sentence, of the way sentences combine to create meaning, coherence, and accomplish purposes" (Gee and Handford 2013, 1). These definitions are, however, very general. In fact, they're only the tip of the iceberg, providing a 'starting point' for further deliberations—the phenomenon of language and its materialization in the form of a text. We should expect nothing more. This is how discourse is described by Jørgensen and Phillips, for instance: "a particular way of talking about and understanding the world (or an aspect of the world)" (Jørgensen and Phillips 2002, 1). Similar sources of it are brought to light in *The SAGE Encyclopedia of Qualitative Research Methods*: "discourse, in the most general sense, is the study of language as it is used in society expressed either through conversations or in documents" (Given 2008, 216). Angermuller et al.'s panorama gives a good idea of the extent of this phenomenon and the difficulty involved in the attempts to describe it further and in more detail.

The trail we will follow further here is provided by Norman Fairclough—he draws attention to a particular (though not very original) context, as it is accompanied by many similar opinions. Yet, he opens the door to a well-founded variant of understanding discourse that elaborates on, somewhat unexpectedly, Boole's vision and brings us closer to the most important topic: knowledge. He writes as follows: "I see discourses as ways of representing aspects of the world – the processes, relations and structures of the material world, the 'mental world' of thoughts, feelings, beliefs and so forth, and the social world. Particular aspects of the world may be represented differently, so we are generally in the position of having to consider the relationship between different discourses. Different discourses are different perspectives on the world, and they are associated with the different relations people have to the world, which in turn depends on their positions in the world, their social and personal identities, and the social relationships in which they stand to other people. Discourses not only represent the world as it is (or rather is seen to be), they are also projective, imaginaries, representing possible worlds which are different from the actual world, and tied in to projects to change the world in particular directions" (Fairclough 2003, 124).

Fairclough's point of view is heavily influenced by Foucault, whom I have covered in Chap. 4. According to Fairclough and Angermuller, as well as other researchers, Foucault is an author who has a major influence on discourse analysis in the social sciences (Fairclough 2003, 2; Jørgensen and Phillips 2002, 12). Following this path is by no means eccentric. Rather, this path leads us along a solid and safe route. Both Paltridge and Fairclough derive its idea from their observation of the general bifurcation of the path followed by discourse analysis. One of its branches goes through the territory of texts, exploring their properties. The other is bolder and allows itself to penetrate deeper, more adventurously into the world. Paltridge says it represents "more socially oriented views of discourse analysis which consider what the text is doing in the social and cultural setting in which it occurs" (Paltridge 2006, 1). Faiclough argues that it is heading toward what he calls "the 'order of discourse', the relatively durable social structuring of language which is itself one element of the relatively durable structuring and networking of social practices" (Fairclough 2003, 3). It is this second, creative avenue of research, which Fairclough bases on Foucault's concept, that directs us toward a reflection on language and discourse that encompasses the problem of knowledge.

This process is fueled by the idea of social constructivism, which makes it possible to interpret the social context that Paltridge and Fairclough speak of as a free (relatively speaking, according to Foucault's teachings) space for the circulation of beliefs, concepts, theories, truths, and so on. According to the critical school of thought, the discourse that accommodates all these elements in some peculiar way, puts them into some specific arrangement, affects situations, knowledge, identities, and social relations (Wodak 2011, 39). According to Crane, who stays true to the field of Knowledge Management, social constructivism considers the social context as its immediate backdrop. Knowledge so understood is no longer any form of objectivity. Instead, it is produced in social interactions (Crane 2016, 83). Jørgensen and Phillips write point-blank that socially constructed knowledge should be expected to support a diversity of discourses, which seem to have found the perfect 'habitat' to exist here: "The struggle between different knowledge claims could be understood and empirically explored as a struggle between different discourses which represent different ways of understanding aspects of the world and construct different identities for speakers (such as 'expert' or 'layperson')" (Jørgensen and Phillips 2002, 2). When it comes to details concerning the unquestionable existence of such a social reality, the authors refer to Finn (Finn 1997). Angermuller et al. have no doubt that discourse, representing a certain reality, descending to the most basic functions of language, does not stop there. Indeed, as they write, "representing the world can also mean constituting it in a certain way" (Angermuller et al. 2014, 6), which is directly related to the context of knowledge. They have no doubt that representation is a process of knowledge creation and at the same time remains involved in its simultaneous coming to being.

We are thus approaching some very important conclusions. Even a very cautious approach to the problem of knowledge in the context of discourse, limited to the simplest assumption that discourse must be based on knowledge, if only as an utterance of that knowledge, sets in motion the context of its emergence. What is noteworthy here is the evolution of the views of Teun van Dijk, an unquestionable authority in discourse studies. Initially, he considers this context in an individualistic, personal, unitary way, focusing on the mental circumstances of knowledge formation (Dijk 1997b). In the years to follow, however, the area he focuses on expands substantially, in line with the opinions mentioned above. He writes as follows: "Discourse presupposes (semantic) situational models of events talks about, as well as (pragmatic) context models of the communicative situation, both construed by the application of general, socially shared knowledge of the epistemic community" (Dijk 2013, 601). However, acknowledging, giving significance to this "socially shared knowledge", which is not "external", transcendent to man, or "objective", produces an image of knowledge that is decentralized, subject to dispersion, variable, numerous, with undetermined internal relations. To put it in a nutshell: it creates a space of multiple discourses that are the realization of different "knowledges", for not even an instance of some superior knowledge.

The Polish language does not have an established, generally approved plural form of the word "*wiedza*" ("knowledge"). It seems that the case of the English language is not that different in this regard either. This is the legacy of viewing knowledge as an idealistic construct: the one and only, the ultimate. But it's a legacy only partially materialized, though one that still has a number of supporters. Without getting into fruitless arguments about who is right or about the truth, let us assume, according to the interpretation offered here, that there is also knowledge that exists in realizations that are numerous, simultaneous, and not necessarily exclusive. Their mutual positioning, their interplay does not let itself be ordered, organized in a uniform manner, for it is time-varying, local. But this is a vision that had already been proposed by Boole. At the time, however, it was depicted as a logical and—in its implementation—rhetorical device. The path that has been traveled since those observations has run through basic reconstructions concerning the imagining of the world and the relations connecting that imagining to the world. It ran through stretches of geometry and mathematics before turning towards philosophy (epistemology). The 'transformation' included in the title is precisely this historical and cognitive process, born essentially at the end of the nineteenth century and continuing till this day after the turbulent and theory-, discovery-, and finding-rich twentieth century. The process in question took place mainly in the territory of physics and mathematics—in a seemingly material reality, one that is real and one that appears to be permanent and indisputable. So the blow to such conviction, which was dealt by the said process, was very painful and heavy. This blow reached also, not surprisingly, the humanities or social thought.

References

Angermuller, Johannes, Dominique Maingueneau, and Ruth Wodak. 2014. *The discourse studies reader: Main currents in theory and analysis*. Amsterdam/Philadelphia: John Benjamins Publishing Company.

Audi, Robert, ed. 1999. *The Cambridge dictionary of philosophy*. 2nd ed. Cambridge: Cambridge University Press.

Badiou, Alain. 2008. *Number and numbers*. 1st ed. Cambridge: Polity.

Boole, George. 1854. *An investigation of the Laws of thought: On which are founded the mathematical theories of logic and probabilities*. London: Walton and Maberly. http://archive.org/details/investigationofl00boolrich

Cheng, Ta-Pei. 2005. *Relativity, gravitation and cosmology: A basic introduction*, Oxford Master Series in Physics. Oxford, UK: Oxford University Press. https://books.google.pl/books?id=sE47I357JjIC

Crane, Lesley. 2016. *Knowledge and discourse matters: Relocating knowledge management's sphere of interest onto language*. Hoboken: Wiley. https://doi.org/10.1002/9781119079316.

Dascălu, Mihai. 2014. Computational discourse analysis. In *Analyzing discourse and text complexity for learning and collaborating: A cognitive approach based on natural language processing*, ed. Mihai Dascălu, 53–77. Cham: Springer International Publishing. https://doi.org/10.1007/978-3-319-03419-5_4.

Davis, Martin. 2000. *The universal computer: The road from Leibniz to Turing*. 1st ed. New York: W. W. Norton & Company.

De Fina, Anna, and Alexandra Georgakopoulou, eds. 2020. *The Cambridge handbook of discourse studies. Cambridge handbooks in language and linguistics*. Cambridge: Cambridge University Press. https://doi.org/10.1017/9781108348195.

De Morgan, Augustus. 1966. In *On the syllogism: And other logical writings*, ed. Peter Heath, 1st ed. London: Routledge & K. Paul.

Dummett, Michael. 1994. *Origins of analytical philosophy*. Harvard University Press.

Fairclough, Norman. 2003. *Analysing discourse: Textual analysis for social research*. London: Routledge.

Finn, Collin. 1997. *Social reality*. London/New York: Routledge. http://archive.org/details/finn-collin-social-reality

Foucault, Michel. 2002. *The order of things: An archaeology of the human sciences*. London: Routledge.

Gee, James Paul, and Michael Handford, eds. 2013. *The Routledge handbook of discourse analysis*. 1st ed. London: Routledge.

George, Rolf, and James van Evra. 2006. The rise of modern logic. In *A companion to philosophical logic*, ed. Dale Jacquette, 35–48. Malden: Blackwell Publishing. https://doi.org/10.1002/9780470996751.ch4.

Given, Lisa M. 2008. *The SAGE encyclopedia of qualitative research methods*. London/Thousand Oaks: SAGE.

Golbeck, Jennifer. 2013. *Analyzing the social web*. Amsterdam: Elsevier.

Harris, Zellig S. 1952. Discourse analysis. *Language* 28 (1): 1–30. https://doi.org/10.2307/409987.

Hyland, Ken, and Brian Paltridge, eds. 2011. *Bloomsbury companion to discourse analysis*. London/New York: Bloomsbury.

Jørgensen, Marianne, and Louise Phillips. 2002. *Discourse analysis as theory and method*. London/Thousand Oaks: SAGE.

Jurafsky, Dan, and James H. Martin. 2009. *Speech and language processing: An introduction to natural language processing, computational linguistics, and speech recognition*. 2nd ed. Upper Saddle River: Prentice Hall.

Kamath, Uday, John Liu, and James Whitaker. 2019. *Deep learning for NLP and speech recognition*. 1st ed. Cham: Springer.

Keynes, John Neville. 1900. *Studies and exercises in formal logic*. 4th ed. London: Macmillan and Company. http://archive.org/details/studiesandexerci029427mbp

Kline, Morris. 1990. *Mathematical thought from ancient to modern times*. Vol. 3. New York: Oxford University Press.

Lyotard, Jean-François. 1983. *Le Différend*. Paris: Les Éditions de Minuit.

———. 1988. *Le Différend*. Trans. G. Van Den Abbeele. Minneapolis: University of Minnesota Press.

Maciąg, Rafał. 2012. *Deus ex machina: nowe media i ich projekt poznawczy*. Kraków: Wydawnictwo Uniwersytetu Jagiellońskiego.

Magala, Sławomir. 2021. *The third enlightenment (or globalizing meritocracies)*. Newcastle upon Tyne: Cambridge Scholars Publisher.

Maus, Heinz. 1962. *A short history of sociology*. New York: Philosophical Library. http://archive.org/details/shorthistoryofso0000maus

Paltridge, Brian. 2006. *Discourse analysis. An introduction*. London/New York: Continuum.

Pêcheux, Michel. 1969. *Analyse Automatique Du Discours*. Paris: Dunod.

Rorty, Richard M., ed. 1992. *The linguistic turn: Essays in philosophical method*. Reprint edition. Chicago: University of Chicago Press.

Schiffrin, Deborah, Deborah Tannen, and Heidi Ehernberger Hamilton. 2001. *The handbook of discourse analysis*. Chichester: Wiley.

Schober, Michael F., David N. Rapp, and M. Anne Britt. 2018. *The Routledge handbook of discourse processes*. 2nd ed. New York: Routledge.

Sriram, Nambiar James. 2000. The Influence of Aristotelian Logic on Boole's Philosophy of Logic: The Reduction of Hypotheticals to Categoricals. In *A Boole Anthology: Recent and Classical Studies in the Logic of George Boole*, ed. Gasser, James, 217–239. Dordrecht: Springer.

Swingewood, Alan. 1991. *A short history of sociological thought*. London: Macmillan Education UK. https://doi.org/10.1007/978-1-349-21642-0_1.

van Dijk, Teun A., ed. 1997a. *Discourse as structure and process. Vol. I. Discourse studies: A multidisciplinary introduction*. London: SAGE.

———. 1997b. The study of discourse. In *Discourse as structure and process. Discourse studies: A multidisciplinary introduction*, ed. Teun A. van Dijk, vol. I, 1–34. London: SAGE.

———. 2013. Discourse and knowledge. In *The Routledge handbook of discourse analysis*, 587–603. London: Routledge.

Weik, Martin H. 2001. Discourse universe. In *Computer science and communications dictionary*, ed. Martin H. Weik, 426–426. Boston: Springer US. https://doi.org/10.1007/1-4020-0613-6_5215.

Wiedemann, Gregor. 2016. *Text mining for qualitative data analysis in the social sciences*. Wiesbaden: Springer Fachmedien Wiesbaden. https://doi.org/10.1007/978-3-658-15309-0.

Wodak, Ruth. 2011. Critical discourse analysis. In *Bloomsbury companion to discourse analysis*, ed. Ken Hyland and Brian Paltridge, 38–53. London/New York: Bloomsbury.

Chapter 10
Poincaré and All

"It is impossible to represent to oneself empty space; all our efforts to imagine a pure space, whence should be excluded the changing images of material objects, can result only in a repre- sentation where vividly colored surfaces, for example, are re- placed by lines of faint coloration, and we can not go to the very end in his way without all vanishing and terminating in nothing- ness. Thence comes the irreducible relativity of space. Whoever speaks of absolute space uses a meaningless phrase" (Poincaré 1913, 413). The quoted words come from a book published in 1908, written by the eminent mathematician Henri Poincaré, without whom mathematics might have not developed as it has (Poincaré 1908). He is the last man, as Torretti summarized what mathematics historians thought about him, who possessed universal knowledge of his discipline and its applications (Torretti 1978, 255). It was this unusual and unique advantage—embracing a field characterized by detail and breadth unavailable to others—that probably let him enter and explore not only the world of earliest and basic findings concerning mathematics or geometry, but also science and the constructs it created. Indeed, Poincaré is also studied as a philosopher of science (Sady 2013).

The words quoted could be the beginning of at least two different stories. First, a story of a brilliant mathematician who does not need to use formulas and numbers to talk about mathematical reasoning. Second, a story of a very long process, during which the sense of presence turns into an increasingly difficult and abstract awareness of the place where this presence occurs. The thinking of such a place has come a long way and has changed over the centuries. We could call this place a space, and then it becomes a kind of separate and special entity. It is also subject, like any concept, to some interpretation. The concept of space is initially close to the everyday experience of a place that is historical and material, tangible. Over time, this experience becomes more and more abstract. The same is true for the rational elaboration of the nature and properties of this place, dissected into many concepts analogous to space—although the latter remains its most general summation in the form of geometry. Geometry is, therefore, a description made using selected means and principles. Poincaré thus also proposed a completely new understanding of geometry.

R. Maciag, *Knowledge as a Tale*, https://doi.org/10.1007/978-3-031-69820-0_10

The stakes involved in developing geometry and its most important concept—space—are very high. Its true value is revealed when it reaches a scale beyond human reach. This is what happens when man observes the cosmos and its spatial relationships. Such a scale cannot fail to address the question of the beginning, and it is at this scale that complete human helplessness and demands for a higher instance—the divine—come to light in their full obviousness. Viewed this way, the concept of space opens an ultimately metaphysical perspective, as it raises the question of the source of the existence of the cosmos and the world. In such a light, space acquires a completely fundamental and transcendent nature. It is therefore quite clear that it becomes the subject of deep thought and many different views that follow. It also arranges itself into a separate story of the evolution of the concept.

Such a perspective also makes the field of deliberation explode far beyond the domain of mathematicians and becomes rather a territory suitable for philosophers. This effect is confirmed by the presence of thinkers such as Descartes and Kant in the discussion. By the force of its vastness and fundamental nature, the problem of space turns out to be an issue that unites the areas of all thought, including bridging the gap that divides the attitudes of the strict type—for instance, the mathematical attitude—with humanistic attitudes—i.e. philosophy—and manages to encompass art at the same time. It appears to be a kind of common—or rather extraterritorial—area that opens up to a bold and multifaceted reflection, which we would today call *transdisciplinary*.

We can find many places that confirm this, which include, for example, the Renaissance revolution that took place in the field of painting, which, in its search for a realistic representation of the world, had to face the completely geometric problem of depicting a three-dimensional world on a two-dimensional canvas plane. Kline, when dealing with the history of mathematics, addresses Alberti's treatise entitled *Della pittura* (painting, image)—which inspired many later authors—at great length. Alberti wrote it in 1435, but it was printed many years after his death, in 1511 (Kline 1972, 232). Kline considers this work to be entirely mathematical and written by a genius. It raises the problem of convergent perspective—a Renaissance invention aiming to solve the geometric problem of three-dimensionality. Painters such as Brunelleschi, Masaccio, and Uccello also turn out to be mathematicians, making a sacrilegious intrusion into a realm which may not necessarily offer the promised reward at all given the increasing knowledge of the spatial relationships that shape and govern the world. In her novel *The Journey of the Book-People*, Olga Tokarczuk describes Uccell's defeat witnessed by his contemporaries. The writer traces its cause to an exaggerated exaltation of the power of mathematical formulas over free imagination. However, Uccello, considered an expert in the domain of perspective, became an inspiration for many master painters who came after him. The idea of perspective, which originates in the school of Florentine artists, continues to develop both in Italy and elsewhere in Europe for the next several hundred years (Andersen 2007, 714).

According to Jöst, the Renaissance brought about the first major breakthrough in the understanding of space. The change was inspired not only by the invention of convergent perspective, but also by the progress in the field of sea voyages. Traveling by sea required appropriate maps, which acted as the representation of a three-dimensional—and spherical (as will soon turn out)—world on the surface of drawings. It became a movement that brought together both theorists and practitioners (to borrow Andersen' distinction) and made use of geometry to analyze a very special aspect of human activity—art. Geometry, which belongs, in fact, to the realm of mathematics and which serves as the basis for art, became somewhat updated, upgraded, refreshed, and brought to the surface. Andersen's very detailed work did not give rise to a great and direct impact of the perspective theory on the development of mathematics and geometry. Jöst sees there some connection with Riemann and Klein, which leads to algebraic geometry in the nineteenth century. However, it is essential to just address the problem of geometrization of the world in connection with certain practical needs (of art). In retrospect, it is the update and revelation of a special geometry that was, at the time, taken for granted and unquestionable. This refers to Euclidean geometry. It dominated the field of mathematics until the nineteenth century. Considered from the perspective of common experience, which appears e.g. as the learning of geometry in school, it continues till this day.

It is somewhat ironic that there is no certainty regarding the circumstances of Euclid's life. Even the dates of his birth and death are not certain. He is said to have lived between 400 and 300 BC in Alexandria, which was ruled by the Ptolemies at the time and developed into a remarkable city, a mecca of poets, philosophers, astronomers, and mathematicians. It was in Alexandria that a center created with research and teaching in mind was established. Dedicated to the Muses, it was known as the Museum. It was also there where a public library with a collection 750,000 volumes was built. The streets of Alexandria were bustling with a cosmopolitan crowd of many nationalities; mechanics flourished, science progressed. Euclid, Kline argues, in ordering Greek mathematics of the earlier, classical period, paved the way for a whole legion of Alexandrian mathematicians, including such prominent figures as Erastotenes, Archimedes, and Heron (Kline 1972, 104). His most important work, *The Elements*, did not make it to modern times in its original form, but it did survive in a great number of copies, which was due to its huge popularity. This work, as proven by the great number of its editions, "is the first substantial source of mathematical knowledge and one that was used by all succeeding generations, it influenced the course of mathematics as no other book has. The very concept of mathematics, the notion of proof, and the logical ordering of theorems were learned by studying it, and its contents determined the course of subsequent thinking" (Kline 1972, 86).

Another circumstance, strongly related to the aforementioned popularity, is of the essence to our argument. The geometry that Euclid describes is in close and direct relation to everyday experience, and is seen simply as a state of the world. *The Elements* consist of 13 books. The first begins with definitions of certain concepts, such as point, line, or surface—which will be used in further reasoning. These are followed by five postulates (e.g. it is possible to draw a line from a point to a point) and five general concepts (e.g. things that are equal to a third thing are also equal to each other, or the whole is greater than the part). The latter were called axioms by Proklos (Kline 1972, 59). They play a critical role to and form the basis of Euclid's further argument—his theorems and proofs, comprising some complete and internally consistent system. This system is produced through deduction, in which conclusions become premises in the steps that follow. At the beginning, there are axioms. They are a kind of initial, first, most thorough or basic assumptions on which further inference is built. Most importantly here, they are also considered some indisputable and unquestionable truths that name certain principles that govern our world. It's hardly surprising that such a belief persists for a very long time. According to Murawski, "in fact, until the end of the 19th century, mathematicians were convinced that axioms and postulates should simply be statements of truth. That is how sentences that describe the real state of affairs appear in mathematical reality" (Murawski 2004, 576). Thus, Euclid's geometry was considered a true story about the world known to man from everyday experience, which confirmed this truthfulness at every possible occasion. Moreover, it continues to confirm it to this day. This geometry is actually the only commonly known interpretation of the spatial relations we experience. No geometry other than Euclidean geometry is taught in elementary schools and in the overwhelming majority of secondary schools. The reality of its formal description adheres directly to the real world so well, so tightly that it is as if glued to it—like a perfect mask adhering to the face after which it has been modeled.

The breakthrough of which Jöst speaks occurs through the growing scale of reflection on geometry, which embraces the world around man in an increasingly bolder manner. Torretti cites Kepler as follows: God always geometrizes (Torretti 1978, 23). Kepler makes no distinction between physics applicable on Earth and beyond, acknowledging geometry as the universal key to the entire universe. In this context, Euclid's constructs take on a new meaning. His triangles and circles find a direct way to materialize in nature; these sets of points must somehow exist. However, the effect of such thinking is quite surprising. Torretti believes that this is how the idea of space, which is not offered explicitly in Euclid's writings, comes to being. He writes: "we may say, therefore, that the assumption that geometry is the basis of physics and that the world is a realization of Euclidean theory implies the existence of Euclidean space" (Torretti 1978, 25). Torretti goes to great lengths to acknowledge this moment as the birth of a contemporary understanding of space, one that can break away from its immediate, material manifestation, and rise to the level of a certain abstraction. This abstraction does not mean, of course, the lack of direct correspondence with the real world—this breakthrough would not come about until the nineteenth century. For Euclid, space did not exist in this advanced form. It turned out, however, that his geometry could *implicitly* and somewhat unconsciously assume such an existence. Actually it must have been so, since it was assumed beyond any doubt that it was real, while at the same time it became increasingly necessary to conclude that the real world without an independent idea of space is simply impossible to accept.

This step is speculative and rhetorical, but expresses a certain ontological conviction. Space emerges as a separate, emancipated, and independent being. The primacy of Euclid's design is by no means disturbed in such circumstances. Sixteenth-century philosophers, such as Giordano Bruno and Tommaso Campanella, and then such powerhouses as Leibniz and Newton in the following centuries, elaborate on and develop this concept—but in different ways. Jöst describes the problem the latter faces in connection with his revolutionary idea, which contradicts previous views. The advocates of these views included, for example, Descartes, who, clearly influenced to some extent by Aristotle, still viewed space as the habitat of directly interacting physical bodies. Newton went a step further, introducing the idea of forces acting remotely, at a distance, allowing dynamic relationships of objects distant from each other, which required the adoption of a concept of new and different phenomena of their connections, of certain forces affecting these objects—while devoid of corporeal form at the same time; forces such as gravity, so obvious today, for instance. This idea is fundamental to the further development of physics, Jöst stresses, because replacing the problem of forces traversing space with the idea of a field leads directly to Einstein's general concept of the interaction between space and matter (Riemann and Jost 2016). Newton speaks of space in absolute terms. By introducing a three-dimensional coordinate system into its geometric interpretation, he makes it the reference system for any physical process. This way, space, still considered in Euclidean terms, eventually becomes sovereign, independent.

Newton makes an invention as well—made independently by Leibniz too. Kline describes this invention as the second greatest work in the history of mathematics after Euclid's concept (Kline 1972, 342)—the differential calculus. The idea of this calculus is formed gradually, but its precise, complete, published explanation is offered exactly by Newton in several works, published many years later in the early eighteenth century. The great significance of this calculus is about its ability to formally capture a phenomenon that has been known since antiquity to be fundamental to the nature of the universe and—at the same time—elusive to mathematical discipline. The movement in question here is a necessary result of forces filling space and makes space time-varying, which produces a certain requisite whole that imagines the world in all its scales. To tame it in the form of mathematical formalism is an effort toward ordering the foundations of this world and, at the same, time complementing its conceptual representation in the form of space.

An even more difficult concept of space is offered by Kant, who, while defending Euclid's concept on the one hand and creating a philosophical justification for Newton's idea on the other, considers space as a form of intuition inherent in the subject observing the world. By taking this step, Jöst writes, he makes space relative, contrary to Newton's own intentions (Riemann and Jost 2016, 17). This leads to viewing geometry as a certain construction, potentially changing with the subject who is the source thereof. In Kant's argumentation, Torretti sees the revolutions that were to come in the near future. Kant, viewing space as a multitude of a priori intuitions, did not imagine it as a mere set of points, but rather a certain continuum. While it must have had some internal relations, and thus some topology, also of intuitive origin, "subject to no other laws than its own" (Torretti 1978, 33), it is completely free. Since this is the case, thanks to this 'freedom' of arbitrarily established laws, even the theorems of classical geometry need not be necessary; "nothing can prevent the understanding [of space – author's note] from developing a variety of alternative geometries (compatible with the prescribed topology), and using them in physics" (Torretti 1978, 33). The field for ideas that transcend the homogeneous, common, 'mandatory' idea of space—and, further, of its geometry—is finally opened.

Until the nineteenth century, the discussion on space and geometry—the science that dealt with space, intensified. But it was not until this century that the real revolution began. Both became the main point of consideration and reflection, and resulted in unusual and very far-reaching ideas, and the number of mathematicians involved in this effort was impressive. Mathematics and geometry turned out to be subject to completely fundamental changes, and eventually paved the way to a situation whose revolutionary nature has already been covered in Chap. 5. This process was very basic nature and had far-reaching consequences, which I have described elsewhere.[1] One of the main problems related to the situation in question was the reconstruction of the idea of space. These events lead directly to the figure and ideas of Henri Poincaré, who was also one of the chief originators of this reconstruction.

These profound changes are also caused by the evolution of the notion of space, which emerges precisely as an interpretive problem. The need to consider the ontological status of space and, at the same time, to have a good understanding of the meaning of its name makes itself forcefully present and proves impossible to circumvent. Before the nineteenth century, as Torretti argues, space was considered an immaterial medium, populated by Euclidean geometric figures consisting of immaterial, intangible points (Torretti 1978, 25). Now, however, with the development of thinking about space, its ambiguity, especially in the context of the material world and man it describes, calls for a question about what it actually is. Necessary questions regarding its ontological status appear, and thus, first of all, questions about whether it is independent of the matter that fills it, or whether it is perhaps a construct of the mind that looks at this matter, as Kant wanted to see it. Some of the greatest philosophers, including physicists such as Newton, Leibniz, Melabranche, Kant, as Torretti continues, agreed on the fundamental properties of space—i.e. that it is continuous, infinite, three-dimensional, homogeneous, and isotropic. But above all—that it fulfills the criteria of the idea of geometry as proposed by Euclid. The last two properties, which are less obvious, meant respectively that space has no distinguished points and no distinguished directions. Such a consistent, extensive concept meant anchoring the idea of space on a 'solid foundation' (Torretti 1978, 25). However, this foundation proves weak and insufficient. It cracks under the pressure of thoughts and imagination gone free.

One can say that the humanist project, which emerges powerfully as an anthropological project in the period between the fourteenth and sixteenth centuries, thus accompanying the transformation of the understanding of space, finally reaches its peak. This project was named 'humanism' by the Bavarian teacher Friedrich

[1] I describe these developments in my book *Digital Transformation. A Story of Knowledge*. Its main thesis directly concerns the aforementioned breakthrough in mathematics. Speaking more precisely, I consider it as the beginning of a new paradigmatic way of knowing, perceiving, and exploiting the world. It is based on digital machines, it is immanently digital, and continues a breakthrough that was highly speculative, but then took over all of civilization.

Immanuel Niethammer only in the nineteenth century, when it finally became possible to summarize it within a distanced and abstract framework. This "educational and cultural program, based on the study of classical Greek and Latin authors" (Kristeller and Randall 1956, 2), is aimed at literature as a unique disposition of man capable of producing worlds equal to those that have hitherto appeared only in books of God. Thus, in a revolutionary way, it defines the human being and their capabilities, in line with the Christian idea, but through an actually definitive and heretical emancipation of this being, which also has a context that is highly important for the story we are dealing with here. In these circumstances, literature appears not as a place to philosophize—and reflect, but rather to create and invent. Man proves able and entitled to pursue and achieve such a feat, while reinforcing their sovereign status in relation to the world—and, necessarily, to God. The culmination of this emancipation comes with Kant and his famous 1785 text *Grundlegung zur Metaphysik der Sitten* (Groundwork of the Metaphysic of Morals), written 15 years after the first lecture of his new philosophy of space and time, in which, as Torretti reminds us, "the ontology of space is, at the same time, an epistemology of geometry" (Torretti 1978, 29). Thus, the image of the world and the presence of man in it comes to form a certain uneasy whole with the world's formal, geometric, mathematical interpretation. One can see here a necessary interdependence between these elements, but this interdependence is governed by man, an instance hitherto not obvious in this context, yet now empowered and equipped with extremely effective tools for interfering with the image of the world.

A remarkable showcase of man's power and, at the same time, an explosion of man's creative power is the work that appeared in the nineteenth century and initiated a profound transformation of mathematics, taking it into a completely new world—a world that is free and unrestricted beyond imagination. This work is the so-called non-Euclidean geometry, with a real breakthrough hidden behind its not very friendly and technical name. Its three main originators, i.e. Friedrich Gauss, Nikolai Lobachevsky, and János Bolyai, decided to take advantage of said freedom, although their ideas, as Klein soberly points out, were based on the work of their predecessors. And the latter two mathematicians, considered the first and most important, owe much to the former. Klein emphasizes that it's important not to overestimate individual contributions, but to look at the broader picture. This is an interesting addition to the image of man freed in their creative pursuits, but surrounded by other people and experiencing the world together with them. This dilemma and dialectic of the processes of recognizing and making use of the world became the main point of consideration in the second half of the twentieth century, and is the subject of extensive discussion in this book.

The starting point of the reasoning of the three mentioned mathematicians appears to be very technical. It draws conclusions from a certain detailed assumption that Euclid introduces into his geometry—it acted as an axiom, i.e. a generally accepted source of reasoning. Except that Euclid believed that this assumption was grounded in truth, while our mathematicians considered it the subject of some arbitrary decision, which is more like a thought experiment than a description of reality. They chose the fifth postulate, concerning parallel lines, which we all must have come across at school. According to this postulate, the angles at which a line intersects other two parallel lines must add up to a certain magnitude (180°). If this is not the case, the lines are not parallel. This postulate has long been problematic because it could not be proven using other axioms from Euclid's system. Thus, in this situation, it seemed entirely possible and reasonable to make the opposite assumption and use it to construct a new system (Kline 1972, 874)—a new geometry.

This simple and rather obvious step, however, meant that the principle of geometry's adjacency to the world was abandoned altogether. Its constructs were treated as free creations, opening the door to mathematics which doesn't need anything but internal consistency. Such mathematics, moreover, was soon found equally useful for describing the world. But what is more important here is yet another consequence of the courage of the three mathematicians: their creation lets us imagine other geometries, and therefore other spaces, which can not only be assumed, but also mathematically constructed and justified. The question about the ontology of one space turns into a question about multiple possible ontologies and spaces. In a world that opens up, the sense of stability of where one is and the search for its justification, for anchorage, is replaced by a certain choice of possible interpretations and justifications. This concept is elaborated on and another idea that reconstructs the field of geometry is proposed by Riemann—which we have dealt with in more detail in Chap. 4. Space turns out to be not so much a specific sovereign entity, but rather a tool, an assumption, an aid. This means entering another—higher—level of abstraction. For such multiplicity is not fiction at all. It exists as if next to the real world, engaging in unusual and complex relations with it, as shown by the twentieth-century physics that builds on such a vision. Absolute, sole, and universal space thus turns out to be an overly simple notion, and its concept—unfounded and therefore devoid of content, as Poincaré, quoted in the introduction, rightly argues.

Space According to Poincaré

Henri Poincaré was a professor of mathematics at the University of Paris, located near the Collège de France, where Michel Foucault gave his lecture on discourse—described here in Chap. 4—decades later. *Genius loci*, that good, protective spirit, seems to have dominated the imagination of both of them, for although they moved in generally different areas of reflection, a certain similarity of thought, a certain correspondence of their basic intuitions about the mechanics of things, emerging from the insights they made and put on paper, made them quite close to each other. Poincaré had a certain quality that made such a connection more likely—he was extraordinarily versatile, which let him reach a level of generality of reflection unavailable to others and often underestimated by those others. As Kline writes, Poincaré published "a vast number of research articles, texts, and popular articles, which covered almost all the basic areas of mathematics and major areas of theoretical physics, electromagnetic theory, dynamics, fluid mechanics, and astronomy" (Kline 1972, 1170). His entrance into the field of science and knowledge theory was thus a natural further step toward a more extensive and complete picture of his own research. And this took the form of a turn towards his own thought and meant addressing the problem of the descriptive and cognitive structures that govern the interpretive and research apparatus that appeared to be highly useful in all those fields that Kline mentions.

Poincaré goes very far along this path. Having yet no support of the social sciences, the only emergent sociology, he turns to psychology, which appears to be gaining popularity at the end of the nineteenth and the beginning of the twentieth centuries. In the period in which Poincaré writes his most important theoretical works, that is, in the first years of the twentieth century, Sigmund Freud publishes his landmark books. Hence the label into which the French mathematician has been 'pigeonholed', making almost every piece of information about him feature the term 'conventionalism'. Such a label, of course, is capable of destroying any thought that is unusual and elusive to common (and—for the same reason—average) reception. Kline argues that Poincaré's best work was a three-volume publication that was gradually published in the last decade of the nineteenth century, titled *Les méthodes nouvelles de la Mécanique Céleste* (New Methods of Celestial Mechanics) (Poincaré 1892). With this work, he joined the ranks of the greatest minds of physics of the modern era, who explored these powerful areas of the world, becoming mentioned along with names such as Copernicus, Kepler, and Newton.

The said three theoretical works are *La Science et l'Hypothèse* (1902), *La Valeur de la science* (1905), and *Science et Méthode* (1908). On the pages of these books, Poincaré considers rather basic issues in mathematics and geometry. They are described and discussed in philosophical language, but one not too difficult. A language that doesn't make the reader get lost or confused, because it is based on reasoning that is consistent and builds constructions gradually, starting from the basics. The intention to keep the right track of thought is not the result of the author's didactic inclination, but rather a persistent attempt to avoid going astray and keeping away from the blind alley that he himself is trying not to end up in. This type of discourse is also based on deeper assumptions about thinking as such, which leads from assumptions and premises along the path of deduction, marked out by Euclid, to further conjectures and conclusions.

This cautious strategy is understandable. Poincaré grapples with very fundamental problems—the kind that underlie the existence of the world and the efforts to describe them; efforts that are inconsistent and uncertain from the very beginning. And this is Poincaré's first observation. He actually finds the triumphant attitude of the Enlightenment and the humdrum declarations of de La Mettrie and de Laplace completely unfounded. Rather, he locks himself within the structures of his own experience and his own thoughts. We can't get beyond our own entrenchment, our own anchorage in the world, which therefore becomes inaccessible—no matter what tools we use. We are determined ultimately by the context in which we find ourselves, which becomes particularly clear in our geometric reasoning: "But if our system is the entire universe, experience is powerless to give information about its absolute position and orientation in space. All that our instruments, however perfected they may be, can tell us will be the state of the various parts of the universe and their mutual distances" (Poincaré 1913, 84).

Poincaré pursues this subjugation to a certain locality and the imprisonment in it with great persistence, going very far, and finds its root in the mind itself. Geometry deals with the study of solid bodies, but not those bodies that surround us as material objects—rather, ideal, unchanging bodies, derived fully from our mind, which, by means of experience, allows us to shape their image. He writes that "the mind has the ability to create symbols and it is in this way that it has created the mathematical continuum, which is only a particular system of symbols. Its capacity is limited only by the necessity to avoid all contradictions, but the mind makes use of it only when experience provides some reason to do so" (Poincaré 1913, 49). This process moves us away from experience, depriving it of the mandate to make statements about the world, because it is approximate, inaccurate, and momentary. As Jost writes, in the case of Poincaré, "geometry, however, is not an empirical science because it is not corrected by sensory experience and is accurate rather than approximate, like all statements obtained empirically" (Riemann and Jost 2016, 142).

Poincaré reveals the reasons why, half a century later, constructivism will eventually overturn the belief in the possibility of knowing the world, even though in its mature form it will add a social context—for example, as the context of the prevailing discursive forms of knowledge. Poincaré also gives a display of his perspicacity and perceptiveness here as well, pointing to language as the main environment where knowledge comes to being. Conducting an insightful comparative analysis of raw fact and scientific fact, he writes point blank: "all the scientist creates in a fact is the language in which he enunciates it" (Poincaré 1913, 332), which follows directly from his conception of the role of the scientist, i.e. a disciplined and self-conscious knower. This time, he builds the foundations of thought which is the subject that Foucault directly deals with and recurs repeatedly in this book. This conclusion is the result of a very extensive analysis of language and its role in science, accompanied by a recollection of similar analyses carried out less than 30 years later by Ludwig Fleck, which were then continued by researchers such as Thomas Kuhn and Bruno Latour. Poincaré, in his humble and relentless quest for sources of knowledge, is among their immediate precursors.

Very far from the world there lies, in hiding, the source of all geometric constructions, defending itself from man's entanglement with the matter of the world around them. There are also certain epistemic standards established explicitly, which later become labeled as conventionalism, in line with a concept that Poincaré himself puts forward. All of the descriptive constructs used by geometry—as well as mathematics—come from man, although they are strongly linked to experience. However, they are not governed by any general law: "The fundamental propositions of geometry, for instance, Euclid's postulate, are only conventions, and it is quite as unreasonable to ask if they are true or false as to ask if the metric system is true or false. Only, these conventions are convenient, and there are certain experiments which prove it to us" (Poincaré 1913, 124). Thus, geometry turns out to be one of many possible creations—constructed, prompted by experience. Just like its individual elements—for example, space. Summarizing Poincaré's idea, Murawski writes that he claimed "that mathematical objects are constructed by the human mind, that there is no field of mathematical knowledge that is independent of the cognizing subject. [...] Poincaré claimed that the axioms and postulates of geometry are (unlike arithmetical theorems!) neither synthetic a priori judgments nor empirical facts. They are conventions and implicit definitions. The choice between different possible conventions is based on experimental facts, but we are free in this choice – the only limitation is to avoid inconsistency" (Murawski 2004, 587).

In such a situation, a situation of almost final removal of constraints and free movement of mathematical constructions in the world, the idea of space explodes in the hands of Poincaré. Firstly, it becomes something different from its physical version, which is a separate entity linked to its more substantial, purely speculative and conceptual realization. Poincaré calls it *continuum amorphe*, whose immaterial, intangible substance—seemingly paradoxical and at the same time making up its core—is "cohesive like rubber, but not rigid like glass" (Torretti 1978, 352). Its idea comes to being as a consequence of the explosion of the potential existence of a theoretical space to be based on the freedom to construct it. It may also be useful for understanding the physical world, which, after all, is accessible conventionally and conveniently, mostly three-dimensionally, while the number of its dimensions may be, in fact, as many as we have "independently excitable nerve-tips", Torretti adds. Poincaré deals with the emerging "multidimensional mathematical continuum", with an unlimited number of dimensions, which is simply a set of individuals, in the second chapter of his *Science and Hypothesis* (Poincaré 1913, 43). Its concept, as Poincaré writes, comes from a generalization of the physical continuum into the world of mathematical creations. If we wish to move one level higher, to the level of the third-degree continuum, we will encounter nothing but pure symbols there. All these continua are, of course, conventions. In other words, they are fictionalized, but with meticulous consistency.

The idea of this unusual and unlimited space appears as early as in the writings of Leibniz, who, according to Kline, in 1697, in his treatise *Characteristica Geometrica*, attempted to create an unusual geometric system, which he called *analysis situs* or *geometry situs* (analysis of place or geometry of place, respectively) (Kline 1972, 1163). It was supposed to describe geometric figures, making use a special set of symbols, making it possible to represent these figures and their properties, and then create combinations thereof—certain specific actions producing results that can be used in subsequent operations. A system similar to that based on quantitative relations, but quite different from those known from the Cartesian system, inherently absolute and limiting the possibility of transformation and change. *Analysis situs* is thus qualitative rather than quantitative, although this case can not necessarily be fully applicable to the general meaning attributed to quantitative approach and qualitative approach. Qualitative geometry considers the general relations between figures, but the description of their similarities does not solely involve counting parameters—especially dimensions in the sense of quantities describing the figures in question. As Dieudonné stresses, "Poincaré was the first who introduced the idea of computing with topological objects, not only with numbers" (Dieudonné 2009, 17). The similarity is of a different kind. Many textbooks try to give an approximation of its meaning, pointing to the procedure of special deformation, which determines the similarity if successful—or lack thereof otherwise.

This way, Poincaré becomes the inventor of topology, which is less interesting here as a fact of the development of mathematics and geometry, but highlights another and very important property of the idea of space brought to surface. Since *analysis situs*, by definition, deprives itself of the traditional approach to dimensions, and actually questions them, it has to produce its own ways of determining position—albeit a position in a very sophisticated and complex space. Torretti explains Poincaré's idea in a very simple and illustrative way, referring to his recursive definition of how to determine the dimensions of a *continuum (continua*—their number is arbitrary, and thus infinite). "Poincaré does not define dimension, but dimension number, that is, a correspondence assigning a characteristic positive integer to each continuum" (Torretti 1978, 353). This number is taken from the observation of the number of continua formed by dividing the initial continuum. This principle is not grounded outside the continuum itself and does not need a universal, external measurement system.

However, there is another aspect to Poincaré's image of space. An aspect that is absolutely fundamental. We have already encountered it, following the mathematician's argumentation and—at the same time—the way in which a structure like space can be understood and used. We are guided into and through this reality by a figure who is 31 years old at a time when their contributions may be of biggest interest to us, i.e. in the year 1885. This figure is a young—but mature—married man, holding a chair in mathematical physics and probability at the University of Paris (Heinzmann and Stump 2017). This is the year that King Oscar II of Sweden, on the occasion of his 60th birthday, announces a mathematical competition. The goal is to find general evidence for the stability of the solar system (Nolte 2010, 37). The challenge is really serious. It can reflect the deep and underlying fears and concerns of the time. Suffice to recall Nietzsche's very famous 1783 text *Über Wahrheit und Lüge im aussermoralischen Sinne* (On Truth and Lies in a Nonmoral Sense). It's a text that contains his famous, deeply pessimistic definition of truth, and the no less pessimistic vision that opens the text. It is a short parable about wise animals, living, however, only for a while on a star suspended in a miserable, faraway part of the universe, which fades quickly and suddenly. In King Oscar's competition, the scale of cosmos develops naturally and necessarily from a seemingly trivial, very well-known and unsolvable physics problem concerning a collision of three and then many bodies, and thus also celestial bodies, which Newton and many others had previously given much thought to. The dramatic nature underlying the scale of the problem seems to be reflected in the events surrounding the competition. Even though Poincaré doesn't come up with a solution, he wins the competition because his contribution is recognized as the most significant. He then publishes his thoughts and reflections in the form of an article. A few years later, he addresses the competition committee to inform them that he was wrong in the conclusions he arrived at and corrects the article, paying for its republishing out of his own pocket. However, by studying the dynamics of bodies in depth, he discovers the opportunity to adopt an entirely new approach, which he develops and completes in his aforementioned great work *Les méthodes nouvelles de la Mécanique Céleste* (New Methods of Celestial Mechanics) (Poincaré 1892). It is there that he does most of the work on the so-called phase space, which we will address next, and "introduces new geometric tools and approaches that have become the driving forces of modern dynamics" (Nolte 2010, 38).

Dynamics! This is the missing piece of the puzzle with the image of the role and capabilities of space invented by Poincaré. Gleick claims that Poincaré viewed topology and dynamical systems as two sides of the same coin (Gleick 1987, 46). This ambivalence and unity is the source of one of the most important constructs of modern physics for describing and analyzing the world. It is covered by Nolte in his textbook *Introduction to Modern Dynamics. Chaos, Networks, Space and Time*: "A unifying viewpoint of physics has emerged, over the past century, that studying the geometric properties of special points and curves within dynamical spaces makes it possible to gain a global view of the dynamical behavior, rather than focusing on individual trajectories. Dynamical spaces can have many different dimensions and many different symmetries. They carry with them names like *configuration space, state space, phase space, and space–time*" (Nolte 2015, 2).

The model of space understood in this way, a model that makes it possible to present motion in a very general sense, as any change of any system in time, is highly universal. It can include, as Nolte writes, both biological phenomena at any level of detail as well as economic processes—such as price fluctuations in the market or the spread of a virus online. In the textbook, he also considers utilizing it to describe neural networks and relativistic physics and gravity. I have found two other non-trivial examples of application of the model in question in an area outside of physics. One is an idea put forward by Byrne and Callaghan, which involves representing social processes as trajectories based on trends observable as narratives (Byrne and Callaghan 2014). The other is Gärdenfors' idea of so-called conceptual spaces—meaning geometric forms of representing knowledge in cognitive processes (Gärdenfors 2000). I described the former in more detail in a 2018 text (Maciag 2018). It refers to the idea of modeling social processes as complex but graspable phenomena through narrative representations, which can then be represented as trajectories in a specially constructed space. Gärdenfors' idea is summarized in the introduction to his later book, written with Frank Zenker, on the practical applications of the concept. It speaks for studying the representation of information at the mental level as "points, vectors and regions in dimensional spaces. On the basis of these structures, similarity relations can be modelled in a natural way in terms of distances in a space" (Zenker and Gärdenfors 2015, 4).

A textbook by Abraham and Marsden provides perhaps the most reliable and solid foundation of our story of Poincaré and his remarkable construction of space, which, strictly speaking, is a proper reactor capable of carrying the dynamic substance of the world. The textbook is titled *Foundations of Mechanics*, with a subtitle that elaborates on it: "A Mathematical Exposition of Classical Mechanics with an Introduction to the Qualitative Theory of Dynamical Systems and Applications to the Three-Body Problem" (Abraham et al. 1980). At the very outset, the textbook emphasizes Poincaré's brilliance revealed in his claim for the resumption of the qualitative approach, which has since become an integral part of mechanics. A qualitative approach in this case means a geometric approach—i.e. one that tries to depict the states of a dynamic system as trajectories in a certain constructed space called 'phase space'. The lines along which this system moves form its phase portrait. The state of the system is therefore not the result of solving the equations describing its motion; in fact, they turn out to be too complex. A simpler—and much more effective—way turns out to be to imagine this movement in a space that comes to being as a combination of all the dimensions that seem relevant and can far exceed the coordinates appearing in the three-dimensional space we have been taught at school in number and character (in line with Euclid's teachings).

Gleick, attempting to offer an outline of the way such a space is constructed, writes as follows: "Every piece of a dynamical system that can move independently is another variable, another degree of freedom. Every degree of freedom requires another dimension in phase space, to make sure that a single point contains enough information to determine the state of the system uniquely" (Gleick 1987, 135). In a sense, one can see (in one's mind's eye) such a geometric creation that is quite different from mathematical formulas typical to the quantitative approach. This fantastic idea allows the study of the geometric phase portrait using geometric (topological) methods instead of going for the usual mathematical analysis, which further substantiates the qualitative nature of the procedure (Abraham et al. 1980). The knowledge of the mover, not only in the trivial three-dimensional space, but in the abstract space composed of any dimensions, is recorded in the path of their movement. Gleick argues that "in phase space the complete state of knowledge about a dynamical system at a single instant in time collapses to a point. […] The history of the system time can be charted by the moving point, tracing its orbit through phase space with the passage of time" (Gleick 1987, 134). We have already answered the question of what these dimensions might be: anything that provides us with information about the system we study and explore. The answer to the question that follows therefrom, i.e. the question of what such a system can be, is similarly general: anything that can be described in these arbitrary dimensions. Dynamic system turns out to be a universal tool used to describe any phenomena that are subject to change over time. Is there anything else we could wish for?

References

Abraham, Ralph, Jerrold E. Marsden, Tudor Ratiu, and Richard Cushman. 1980. *Foundations of mechanics*. 2nd ed. Reading: Addison-Wesley Publishing Company.

Andersen, Kristi. 2007. *The geometry of an art*. New York: Springer. https://doi.org/10.1007/978-0-387-48946-9.

Byrne, David S., and Gill Callaghan. 2014. *Complexity theory and the social sciences: The state of the art*. New York: Routledge, Taylor & Francis Group.

Dieudonné, Jean. 2009. *A history of algebraic and differential topology, 1900–1960*. Boston: Birkhäuser Boston. https://doi.org/10.1007/978-0-8176-4907-4.

Gärdenfors, Peter. 2000. *Conceptual spaces: The geometry of thought*. Cambridge, MA: MIT Press.

Gleick, James L. 1987. *Chaos: Making a new science*. New York: Viking.

Heinzmann, Gerhard, and David Stump. 2017. Henri Poincaré. In *The Stanford encyclopedia of philosophy*, edited by Edward N. Zalta, Metaphysics Research Lab, Stanford University. https://plato.stanford.edu/archives/win2017/entries/poincare/

Kline, Morris. 1972. *Mathematical thought from Ancient to Modern times*. 1st ed. New York: Oxford University Press.

Kristeller, Paul Oskar, and John Herman Randall. 1956. General introduction. In *The renaissance philosophy of man: Petrarca, Valla, Ficino, Pico, Pomponazzi, Vives*, ed. Ernst Cassirer, Paul Oskar Kristeller, and John Herman Randall, 1–20. Chicago/London: University of Chicago Press.

Maciag, Rafal. 2018. The analysis of the internet development based on the complex model of the discursive space. *Information* 9 (1): 7. https://doi.org/10.3390/info9010007.

Murawski, Roman. 2004. Mathematical knowledge. In *Handbook of epistemology*, ed. Ilkka Niiniluoto, Matti Sintonen, and Jan Woleński, 571–606. Dordrecht: Springer Netherlands. https://doi.org/10.1007/978-1-4020-1986-9_16.

Nolte, David D. 2010. The tangled tale of phase space. *Physics Today Physics Today* 63 (4): 33–38.

———. 2015. *Introduction to modern dynamics: Chaos, networks, space and time*. 1st ed. Oxford/New York: Oxford University Press.

Poincaré, Henri. 1892. *Les Méthodes Nouvelles de La Mécanique Céleste*. Vol. I–III. 3 vols. Paris: Gauthier-Villars et Fils. http://www.bibliotekacyfrowa.pl/dlibra/docmetadata?id=109319

———. 1908. *Science et Méthode*. Paris: Ernest Flammarion.

———. 1913. *The foundations of science: Science and hypothesis, the value of science, science and method*. New York/Garrison: The Science Press. http://archive.org/details/foundationsscie01poingoog

Riemann, Bernhard, and Jürgen Jost. 2016. Historical introduction. In *On the hypotheses which lie at the bases of geometry. Classic texts in the sciences*, ed. Jürgen Jost, 9–28. Cham: Springer International Publishing. https://doi.org/10.1007/978-3-319-26042-6.

Sady, Wojciech. 2013. *Spór o Racjonalność Naukową. Od Poincarego Do Laudana*. Toruń: Wydawnictwo Naukowe Uniwersytetu Mikołaja Kopernika.

Torretti, Roberto. 1978. *Philosophy of geometry from Riemann to Poincaré*. Dordrecht: Springer Netherlands. https://doi.org/10.1007/978-94-009-9909-1.

Zenker, Frank, and Peter Gärdenfors, eds. 2015. *Applications of conceptual spaces*. Cham: Springer International Publishing.

Chapter 11
A Discursive Space

This chapter offers a summary of the discussions and deductions contained in the previous chapters and puts forward some rather far-reaching but legitimate conclusions, wrapping up the whole argumentation. The previous chapters of the book try to depict the nature of the emerging elements of this argumentation, which, put together into a certain whole, lead to the construct of discursive space. This construct was invented as the model of dynamic knowledge, and its principle is quite simple. The main element of a discursive space is, as one might guess, discourse, understood as the articulation and retention of knowledge. Discourse is a linguistic phenomenon, which effectively eliminates any doubt as to how it exists. This is supported by a rich body of literature, discussed mainly in Chap. 6. Discourse is assumed to exist, although its nature is not simple. It was described in much detail and with much insight by Michel Foucault. There has been no better description of it so far, and most of the later reflections are based on his idea. I have covered it quite thoroughly in Chap. 4.

However, before we get to the aforementioned conclusions, and before the idea of building a model of discursive space is described, it seems necessary to offer a brief introduction regarding the most general assumptions, which have been presented elsewhere and will only be recalled here (Maciag 2018a, b, 2020, 2022). Thus, in order to reinforce the ontological context of discursive space, and—more specifically—to give a convincing interpretation of the embedding of discourse in the world, I have resorted to Armstrong (1997), who addresses the matter of existence of the world at a fundamental level, following the trails blazed by Russell—and especially by Wittgenstein. Armstrong uses concepts borrowed from both of them, which is easy to understand because of the mutual influence these two philosophers had on each other. The construction of the world as proposed by Armstrong can be depicted as an image of a moving and changing tangle of elements, which he referred to—after Russell—as *particulars*, which seem significant only as an internally connected whole. The nature of this whole is based on a certain state of affairs—not a set of material objects.

R. Maciag, *Knowledge as a Tale*, https://doi.org/10.1007/978-3-031-69820-0_11

The notion of '*the state of affairs*' is fundamental to Wittgenstein, who uses it to represent the way language connects with and forms a relationship with the world[1] in his *Tractatus Logico-Philosophicus*. I tried to explain this relationship, very deep and complex, in a text introducing the idea of discursive space (Maciag 2018a). It acts as a rather effective justification for the existence of a relationship between language and the world, supported by the authority of Wittgenstein. In this simple manner, the relationship not only appears to be clear, but also gains a metaphysical dimension, which opens the imagination to more advanced ways in which language exists in the context of the world. The term 'more advanced' shall be understood here as the introduction of a level of reflection that rises above pure linguistics, but attempts to reach the epistemological level required to address the problem of knowledge. This is the level of reasoning proposed by Foucault, who justified its connection to the level of social phenomena in many of his essential works.

Armstrong becomes a contemporary continuator of Russell's and Wittgenstein's reasoning, providing descriptive tools to help organize the aforementioned level of description. He gives the concept of the state of affairs a very precise shape: "The general structure of states of affairs will be argued to be this. A state of affairs exists if and only if a particular (at a later point to be dubbed a *thin particular*[2]) has a property or, instead, a relationship holds between two or more particulars. Each state of affairs, and each constituent of each state of affairs, meaning by their constituents the particulars, properties, relations and, in the case of higher-order states of affairs, lower-order states of affairs, is a contingent existent. The properties and the relations are *universals*,[3] not *particulars*. The relations are all external relations" (Armstrong 1997, 1).

Armstrong was also very helpful in formulating a direct definition of the nature of the combination of language and reality of the world, reduced to Russell's simple entities, these *particulars*. This combination, this connection can be described by means of a relation of supervenience, which Armstrong defines as follows: "We shall say that entity Q supervenes upon entity P if and only if it is impossible that P should exist and Q not exist, where P is possible" (Armstrong 1997, 11). Explaining this definition, he adds: "supervenience in my sense amounts to entity P *entailing* the entity Q, but with the *entailment* restricted to the cases where P is possible" (Armstrong 1997, 11).

Thus, this relation defines a certain inevitability of the interconnectedness of two entities, concerning the fundamental property of existence. It actually seemed to describe very well the interdependence of language and the world—a description in which Wittgenstein was so dramatically entangled. Filling in Armstrong's general pattern with the content of the opposition studied here, it can be said that the world remains in a relation of supervenience with language if—and only if—it is not

[1] Which is called, more precisely, the totality of facts (not things)—after Wittgenstein.

[2] Which can be translated, given the context, as a diluted or dispersed element.

[3] Another concept taken from Russell, for whom *universals* are predicates describing *particulars*, such as diversity, causality, father, white, for example (Russell 2003, 95).

possible for language to exist when the world does not exist if language is possible. To take it further, following the supplementation, this supervenience would amount, in our case, to the world entailing the existence of language, but only if the world itself is possible. Language thus has a clearly defined boundary.

The use of the idea of supervenience introduces two very serious elements of reasoning. First, it makes it possible to think of this relationship as a process by bringing to light and updating the issue of time. The relationship between language and the world takes shape in the domain of time, which directly follows from the fact that the world is temporal in nature. Language constantly follows the world, with no hope of succeeding. Second, supervenience is a relation with a very high degree of generality and does not determine anything more than existence. This generality opens the door to all kinds of possible interpretations of the ways in which language and its creations exist, but in a rather strict dependence on the world. We have already named one feature of this world as important and influencing language through supervenience. This feature is temporality. The second such ontological feature is the complexity of the world, which must be inherited by the language by virtue of the supervenient relationship. It is the realization of this relationship, brought by the relationship of supervenience, which Armstrong calls *'entailing'*. Both features of the world—i.e. its temporality and complexity—are considered here to be uncontroversial and even trivial, although very important and carrying further implications.

The reference to Armstrong, and through him to Russell and Wittgenstein, brings about yet another important consequence, arising from the search for an answer to the possible question of the modus of existence of this space. It is a generalization that leads to unification. The level of considerations adopted and the conclusions drawn therefrom—especially the definition of states of affairs, allow for a certain ontological unification in that linguistic phenomena—including discourse—are not excluded from the area of the states in question. Thus, discourse, which is a phenomenon of language, viewed as an autonomous phenomenon of retention and articulation of knowledge, remains at the same time part of the world of facts and a representation (by virtue of the relation of supervenience) of a fragment of it. This is a very basic assumption that translates immediately into the way knowledge is perceived. This is because it means that knowledge is part of the world (of facts). Thus, a system such as language is no longer isolated from a system such as the world. This isolation was considered temporary. A similar line of thought opened the part of the book where I looked closer into Gettier's ideas. For this reason, the phenomenon of the subject of knowledge, that is, the one who knows, the knower, needs to be reinterpreted, reinvented, so to speak.

We came across the problem of the subject of knowledge (owner, possessor) in Chap. 1, which took shape also in response to a three-page article by Gettier, proving the inconsistency of the definition of knowledge offered by Plato, linking this knowledge to its inherent, ever-present possessor—man. It develops further into a story about the levels at which speaking takes place, each time building the author's instance differently. This result is perfectly evident in the interpretations of the mode of existence of a text as cited there, which show how ultimately the subject—the one who speaks—escapes accuracy, hiding in the possible levels of interpretation. The following chapters describe the peculiar evolution of the concept of knowledge, which can be summarized as its gradual emancipation. This evolution, on the one hand, shows the subjective nature of knowledge, resulting from the specific nature of the world of the individual who formulates it. On the other hand, it points to the process of knowledge acquiring the properties of a separate instance. In the latter case, it is presented with increasing consistency as a separate phenomenon, sovereign enough to become the basis of social (and even business) processes. Knowledge turns out to be entangled in various ways in this kind of context, which can be endogenous, set upon on the specificity of the very process of its disciplined accumulation. This is the case, for instance, with science according to Fleck's interpretation. It can also be exogenous, when it results from deeper, historical political processes—those that e.g. govern both the distribution of power and the very basic economic matters as described by Bell and others.

The two manners of presenting knowledge seem to be opposite. One leads to the dispersion of knowledge and to the creation of diverse instances of it, while the other is more about imagining it as a homogeneous and separate resource. However, both are somewhat complementary. The very general economic and systemic processes reorganizing the position of knowledge in the world of society and the state are anchored in the simultaneous modification of knowledge production processes, the specificity of which is studied by the sociology of knowledge and by such researchers as Mannheim, Scheller, and Latour. Their parallelism is striking and calls for interpretation. Both lead to the abolition of the notion of knowledge as man's metaphysical experience that arises from their interaction with the world, perhaps most poignantly formulated by Descartes as the dualism of *res extensa* and *res cogitans*. Knowledge experienced this way is ultimate, definitive, and expresses truth. It is present simultaneously on both sides of the aforementioned alternative. More: it is the means of its existence. The reconstruction of the method of interpreting the phenomena of the world, which makes a dramatic appearance at the turn of the nineteenth and twentieth centuries, transforms this configuration completely. First, it changes man's position, stripping them of their metaphysical status as an equal partner in the world and thrusting them into the depths of social constraints. Then, it gives them the right to freely formulate judgments about this world, thus devaluing the world's previous status as the ultimate arbiter.

The said reconfiguration is perfectly visible within the field of the theoretical support for natural science research, which is mathematics at the end of the nineteenth century. But then it also moves further, stepping into the field of philosophy and social sciences. Its effects, on the other hand, quickly appear both in the space of speculative reflection and in the area of purely practical applications—business. An example of the former is Lyotard's concept of a full and complete analysis of the dynamics of social knowledge as free and historically variable narratives; of the latter—the use of knowledge in business processes. Interestingly, the latter area seems very fortunate and fruitful as a field of experience for the use of knowledge that is actually completely reified, reduced to the level of pure utility, which is sought with great scrupulosity and ingenuity. Knowledge Management (KM) treats the phenomenon of "commodification and reification of knowledge" as a significant element of traditional approaches within the discipline (Crane 2016, 38). I describe these problems in Chap. 5.

The most far-reaching conclusions regarding knowledge are offered by Foucault, who links it explicitly to language, organizing and clarifying the reconfiguration just described in one go, so to speak. To this end, he resorts to some redefinition of the existing concept of discourse and designates it as a phenomenon of knowledge. This is the fullest and most explicit identification, which, as Deleuze argues, grasps a certain pre-establishment dynamic from which words and objects emerge. In order to convey its nature, or—more precisely—its complexity, Deleuze uses the concept of multiplicity, which can be understood as manifold—this concept unites mathematical and philosophical reflections in an extraordinary manner, once again revealing the natural affinity between the two spheres in these reflections. This affinity is also present in the concept of discourse, appearing as the idea of a certain isolated and artificial world subject to analysis and discussion, which Boole refers to as the universe of discourse. I discuss this construct and its consequences in Chap. 6. Mathematicality appears here as analytical accuracy and expediency, as well as legitimate arbitrariness. Such features also push him towards the logical calculus, which appears as an attempt to link the real world with formal actions. The universe of discourse can be viewed as an area where such a connection eventually becomes possible, with a thinking—i.e. also calculating—mind becoming its starting point.

In this context, discourse is the reasoning produced, which becomes as if a record made by means of language, the natural matter of thought. It thus brings together two separate spheres: the world (which appears through reasoning) and language (names, descriptive terms) in an unambiguous cognitive context. This way, it provides a ready-made tool for the abovementioned reconfiguration and becomes, when Foucault recalls it, a readily applied construct. At the same time, it puts knowledge in a new place, and knowledge must now appear as a result of the encounter with the world—but this time it is no longer an encounter of the human (mind) with the world, but an encounter of language with the world, a world specially prepared for this encounter. The effects of this approach are as follows: a multiplicity of universes of discourse is possible and allowed, which results in a multiplicity of their arbitrary constructions. The viewed world becomes fragmented, and this fragmentation is never limited in the ways it occurs. There are certain appropriate and adapted concepts in place in each of the worlds so constructed. The universalistic principle of the coherence of these worlds is rejected in favor of internal coherence only. The focus shifts from the subject (human) to the system of concepts and notions—language. Foucault's contribution involves a change in perspective. He views every statement as a potential realization of some universe of discourse. In other words, he sees the arbitrariness of a cognitive decision as an immanent feature of every statement. Thus, the entirety of the statements make up the numerous and not necessarily realized universes in which they themselves take shape as different discourses. Superimposing this image on the social reality of speech leads to the revelation of the world of discourse and the roles that this discourse performs: knowledge and power.[4]

The reasoning just presented justifies the perception of the discourse of articulation/retention of knowledge. It shows the inevitable historical process during which the idea of the phenomenon of discourse emerges against the background of the evolution of the understanding of the phenomenon of knowledge, making it possible to link the two phenomena together. And this is done in two steps. First, through the removal of the human as the sole subject of knowledge combined with its objectification, and second, through the emergence of the construct of discourse and its combination with knowledge. Both of these steps are, of course, the result of long and complex interpretive processes. In this sense, the story of discourse as knowledge is self-contradictory. It is also a story about itself, which is necessary because this story is also some type of knowledge.

The nature and course of these processes also make it possible to identify certain characteristics, certain qualities of this knowledge, resulting from its 'discursivity', so to speak. It is dispersed, and there are no general rules for this dispersion. Discourses may or may not overlap, complement each other or remain disconnected, etc. Since there is no instance controlling the reality of discourses drawn in this way, it cannot be assumed that certain discourses are more legitimate than others. According to Foucault's ideas, such temptations take shape as political gestures and are the result of the exercise of power rather than substantive evaluation. This is true in particular for the application of the criterion of truth in the classical sense (i.e. regarding correspondence). Since discourses are the outcome of an arbitrary approach to the fragments of the world they deal with and apply names and descriptions in their own way, it is not expected that they must meet such a strict criterion. This is what the history of science teaches us, ultimately summed up by Popper as a place of possible falsification only. In the world of discourses, the criterion of truth is not immanent in the definition of knowledge, as is the case with its definition offered by Plato.

The most important feature of knowledge represented by a discourse is its object-oriented, reified nature, the properties of which are determined by the discourse. Since discourses inherit the complexity of the world, it is necessary to apply tools specific to complex systems to them. This property makes it possible to introduce and adopt a quantitative approach to the mainly qualitative manner of argumentation conducted here so far. This is done through the observations already made and offered here, which depict discourse as an entity-sovereign phenomenon that develops over time. These observations are enough to create a phase portrait of this dynamic system, which means representing it as a chart in an appropriately selected space. Each dimension of this space, specific to a given discourse, should make it possible, *ex definitione*, to illustrate the value of the parameter for which it is a scale. The number of these dimensions is arbitrary and determined by the discourse in

[4] By the way, this process is deposited in history, causing the emergence of certain dominant knowledge formations, which Foucault calls *epistemes*.

question. What it means is that it captures all the considered, changing properties of the discourse, acting as degrees of freedom of the dynamic system which this discourse is. Since the chart of the corresponding values in all dimensions is time-dependent, it ultimately takes the form of a trajectory traversed by the discourse in a space made up of these dimensions, representing the state of that discourse at each of its points. A sample chart of this kind has been made for a discourse on the Internet (Maciag 2018b). Since we treat discourse as the articulation and retention of knowledge, such a chart reflects the state of knowledge.

Discourses, and therefore "knowledges",[5] can be numerous. And indeed they are, as proven by the discourse studies described here. This perfectly illustrates and explains the current multiplicity of different types of beliefs, which underlie numerous—often contradictory or conflicting—instances of knowledge, but which serve as the basis for decisions made and actions taken. This kind of pragmatic approach proves the irrelevancy of the applied concept of truth in the space of knowledge. This irrelevancy is a fact of life, although from an ethical perspective, it may give rise to major and just objections. It is also a very serious and well-known problem, which became the subject of a famous discussion between Lyotard and Habermas in the early 1980s. One of the topics raised in this discussion was the fundamental difference in the judgment and perception of the social situation in which humans are. Lyotard viewed it in terms of a constant dissent, dispute (*le différend*), a notion he included in the title of his book (Lyotard 1988). Habermas, in turn, advocated for the possibility of abolishing dissent through an appropriate communicative situation (Habermas 1981). This problem was part of a broader debate, held by the aforementioned philosophers, regarding the conflict between the two formations they represent—modernism and postmodernism. This dispute, which is nearly half a century old, seems to be a thing of the past, but the problem of dissent has been intensifying with the advancement of the Internet and its communication tools, such as Social Network Sites (SNS), Facebook (boyd and Ellison 2008). This observation is verified and confirmed by the emergence of big social movements centered around unconfirmed or at least debatable views and based on extensive social influence—such movements include the anti-vaccination movement, the flat earthers, or the QAnon. Their presence and impact are founded on various types of beliefs, considered true by the members and supporters of these movements. According to the proposed model of discursive space, they belong to the forms of knowledge that appear through discourse and should be able to be presented through this model.

I have described the basics of how to model dynamic systems using phase space as well as the development of this type of thinking about the world and how to represent it in Chap. 7. There is no doubt that applying this approach to a phenomenon such as discourse seems justified. It involves moving to a more abstract and general level of representing phenomena, although at no point does it lose touch with the real world. The only thing that changes is the type of representation, transforming

[5] The Polish language does not use the concept of knowledge in the plural, which seems to be a legacy of the idea of representing knowledge as one, hence its certain awkwardness in this form, however necessary it may be here.

into one that is less direct and ultimately anchored by a much more extensive and largely arbitrary set of parameters, which further become the basis of the dimensions in which they are reflected. This forms a space where we can 'see' the change of this system, this arrangement over time. The key to building this space is this set of sizes, which makes it possible to create the right dimensions. The only problem that remains to be solved in this situation is that of these parameters by which discourse can be captured. The answer to this question is offered by the very expansive field of discourse research, which also uses increasingly sophisticated, advanced, and mature IT tools.[6] The field also provides convenient operationalizations of discourse, essential to experimental research.

Capturing knowledge in a spatial model also allows for further development. This can take place following two paths, so to speak. One path involves moving beyond the notion of space to the more general construct of manifold. According to Riemann's intention, it is an extension and generalization of the idea of space, and, at the same time—as Husserl points out, it can be understood as a peculiar, purely axiomatic, arbitrary set; a set that is relevant to reasoning, abstract, but geometrically interpretable. The other path makes use of the ontological unification mentioned in the introduction, in which the distinction between discourse and the world is abolished. Both paths make it possible to significantly broaden the knowledge perspective—and both seem highly promising. They also serve as a kind of culmination of the argument provided in this chapter and book.

The idea of manifold was discussed in Chap. 4, when discussing Foucault's book, which Deleuze referred to as a step towards the realization of this idea. This somewhat surprising statement concerns a construct primarily associated with advanced mathematics, yet proving to be a potentially useful tool. It is not necessarily very difficult to justify this possibility. In fact, its application is supported by the intention that guided its creator, Riemann, when he was looking for a certain conceptual interpretive figure that would go beyond the already available set, ending with space as the most advanced structure. And this happens at a time when mathematics and geometry free themselves from the obligation to represent the world directly and move on further, towards interpreting it, which requires a system of concepts detached from this world, as if higher-tier and more general. The fruit borne out of such thinking is geometries unknown from experience, although possible and, as it turned out later on, no less 'real', although utilized by much more advanced areas of physics. An analogous transgression underlies the establishment of the manifold, which is, of course, a certain multitude of entities—but a multitude contained in any number of variously appearing dimensions. Thus, in line with the reasoning adopted here, it is more a set of those dimensions which are used, selectively and according to an additional possible unknown order, as the building blocks of the spaces comprising this manifold.

[6]This means, for example, the so-called CAQDAS (Computer-Assisted Qualitative Data Analyses Software) (Fielding and Lee 1998).

This is how one can extrapolate the idea of discursive space, which can include the trajectory of a single discourse, i.e. the articulation and retention of knowledge, because it is based on a single set of dimensions, recovered as if from observing how this discourse manifests itself (in statements) over time. The underlying assumption is that we are dealing with a great number of such discourses, which, as social products, are arbitrary and very diverse. There would be no reason to combine them into a single whole if it were not for the fact that they are knowledge, or more precisely different 'knowledges',[7] connected to each other in social practice—and not just existing along each other. It's actually the opposite: they compete with each other, complement each other, oppose each other, and enter into many more relations, ultimately forming a certain continuum of knowledge. In order to grasp this very continuum and its dynamics, it is necessary to have a structure higher and more general than space, available in the abstract geometric toolkit—manifold. Encompassing all theoretical discourses in one such construct makes it possible to build a theoretical model of all discursive knowledge, although it also makes the creation of more local manifolds achievable, which is more promising for practical reasons.

The other type of extrapolation of discursive space as knowledge space has been set upon a slightly different foundation. This time, its reach extends beyond discourse, which is treated as one of the possibilities for the articulation and retention of knowledge. There are two separate ways where such a possibility comes into being. One concerns the reconstructed figure of the possessor of knowledge, who is not necessarily a human being—such an assumption already directly includes discourse, which, as a linguistic and—therefore—social creation, although materialized through statements made by people, is not coupled to any specific individual. Moving further along this line of thought, one can already give examples of knowledge that is not human in nature at all. Such examples include biological processes that can be interpreted as computational processes with their own knowledge (Horsman et al. 2017) or artificial cognitive systems (Burgin 2015; Dodig-Crnkovic and von Haugwitz 2017). In the latter cases, the severance of the connection between knowledge and the human is complete and final—and its effects are not yet thought through.

The other way for this possibility to be is opened through the ontological unification discussed at the beginning of the book, introduced through Armstrong's concept of the *world of affairs*—whose elements *(particulars)* are determined only by their mutual relationships. This concept is based, as already mentioned, on the ideas of Russell and Wittgenstein. Its maximum assumed generality and simplicity of elements make the world an organized whole. This whole includes all the parts, and therefore also the discourses and their structures. We should bear in mind, however, that even though discursive space is a speculative construct, the dynamic system of

[7] This way, the fact of their sovereign, individual, and fully legitimate existence is emphasized, which is then lost when terms such as 'instances of knowledge', 'types of knowledge', 'forms of knowledge' and the like appear. They all assume the existence of a single knowledge—a knowledge which they are manifestations of.

discourse is real by definition—although specific, and belongs to the world of facts on par with everything else, functioning as one of its fragments. Multiple discourses, realizing a variety of knowledges, can—and do—include other discourses (and other knowledges), producing a complex structure.

The latter sentence delineates the area of the necessary research efforts, which also need to explore further territories. Indeed, a discursive structure needs to be expanded to include the non-discursive protocols of knowledge articulation and retention, brought about by the activity of non-human types of possessors and acquirers of knowledge. This type of structure requires a separate study. Their more obvious examples, which have appeared a while ago, open the door to the field (in fact, it can be much broader) to encompass any of the materialized articulations and retentions of knowledges that we deal with, for example, in the case of DNA, which are well-defined portions of the energy of the bonds connecting atoms in a molecule or the characteristic pattern of sand dunes in the Sahara desert. Each of them is the result of the existence of a system that exploits a selected and—from its point of view—necessary fragment of the world of facts in any configuration. In such a landscape, the dispersion of knowledge becomes inevitable due to the fundamental organization of this world, in which a given system interacts with its environment through knowledge drawn for a particularistic and selfish purpose—for example, as a local need that sustains the existence of a single system, e.g. a social group centered around a belief shared by its members.[8]

The situation of discursive knowledge changes as the perspective broadens. This is because discourse is only a special case of the existence of knowledge and fulfills the needs of a certain system that was originally limited to a single individual, but turned out to be related to larger—social—structures, i.e. users of various discourses (knowledges). The physical means of conveying the sense and meaning of such a variety of knowledge is a statement. However, this means can take any other form for other knowledges, appearing in other systems and specific to them. Knowledge thus turns out to be a real and universal structure that pervades the world of facts—and perhaps also organizes and arranges it. My suggestion is to call the unit of such knowledge a *gnosem* (after *gnosis*, Greek for 'knowledge'). Thus, this knowledge, which is a great, virtually infinite and complex, dynamic manifold of *gnosems*, is the entire, complete knowledge of the world (of facts).

References

Armstrong, David Malet. 1997. *A world of states of affairs*, Cambridge studies in philosophy. Cambridge: Cambridge University Press.

boyd, Danah M., and Nicole B. Ellison. 2008. Social network sites: Definition, history and scholarschip. *Journal of Computer-Mediated Communication* 13: 211.

[8] This image is inspired by Niklas Luchmann's construct (Luhmann 1990), which I have discussed in more detail in (Maciag 2018a).

Burgin, Mark. 2015. *Theory of knowledge: Structures and processes*, World scientific series in information studies. Vol. 5. Singapore: World Scientifc Publishing Co. Pte. Ltd.

Crane, Lesley. 2016. *Knowledge and discourse matters: Relocating knowledge management's sphere of interest onto language*. Hoboken: Wiley. https://doi.org/10.1002/9781119079316.

Dodig-Crnkovic, Gordana, and Rickard von Haugwitz. 2017. Reality construction in cognitive agents through processes of info-computation. In *Representation and reality in humans, other living organisms and intelligent machines*, ed. Gordana Dodig-Crnkovic and Raffaela Giovagnoli, 211–232. Cham: Springer International Publishing. https://doi.org/10.1007/978-3-319-43784-2_10.

Fielding, Nigel G., and Raymond Lee. 1998. *Computer analysis and qualitative research*. 1st ed. London: Sage.

Habermas, Jürgen. 1981. *Theorie Des Kommunikativen Handelns. Band 1: Handlungsrationalität Und Gesellschaftliche Rationalisierung. Band 2: Zur Kritik Der Funktionalistischen Vernunft*. Frankfurt am Main: Suhrkamp.

Horsman, Dominic, Viv Kendon, Susan Stepney, and J.P.W. Young. 2017. Abstraction and representation in living organisms: When does a biological system compute? In *Representation and reality in humans, other living organisms and intelligent machines*, ed. Gordana Dodig-Crnkovic and Raffaela Giovagnoli, 91–116. Cham: Springer International Publishing. https://doi.org/10.1007/978-3-319-43784-2_6.

Luhmann, Niklas. 1990. *Theorie der Gesellschaft*. Frankfurt am Main: Suhrkamp.

Lyotard, Jean-François. 1988. *Le Différend* (trans: Van Den Abbeele, G.). Minneapolis: University of Minnesota Press.

Maciag, Rafal. 2018a. Discursive space and its consequences for understanding knowledge and information. *Philosophies* 3 (4): 34. https://doi.org/10.3390/philosophies3040034.

———. 2018b. The analysis of the internet development based on the complex model of the discursive space. *Information* 9 (1): 7. https://doi.org/10.3390/info9010007.

———. 2020. Ontological basis of knowledge in the theory of discursive space and its consequences. *Proceedings* 47 (1): 11. https://doi.org/10.3390/proceedings47010011.

———. 2022. Theory of knowledge based on the idea of the discursive space. *Philosophies* 7 (4): 72. https://doi.org/10.3390/philosophies7040072.

Russell, Bertrand. 2003. Analytic realism. In *Russell on metaphysics. Selections from the writings of Bertrand Russell*, ed. Stephen Mumford, 1st ed., 91–96. London: Routledge.

Epilogue

The epilogue to the story about knowledge was added by reality in its own, surprising, and fundamental way. This epilogue is also the beginning of a completely new adventure for humanity. This adventure is called great language models, and the pathos with which we announce them here is completely justified.

Large language models are digital technologies that can generate intelligible text. Its internal consistency and adequacy in relation to the context are equal to those of humans. This text is created as an element of one of many communication situations carried out using text, such as dialogue, answering questions, summarizing, searching, and others. The generated text can be considered artificial in the sense that it was created by a complex algorithm. This, however, does not change the fact that an artificial text has an autonomous character and is a separate and fully fledged linguistic event. This fact makes an artificial text a carrier of certain knowledge, which means that it not only belongs to the story of knowledge, but due to its surprising, unexpected, and complex nature, it certainly opens a new chapter of it.

The proof of the autonomy and fullness of this text comes from a reflection that arose many hundreds of years earlier and was born as an interpretation and analysis of other texts perceived as non-human religious texts carrying a message from the deity himself, to whom its authorship was attributed. Insight into these mysteries was initially strictly reserved, but then, especially during the Reformation, it became common and even required. Hermeneutics, whose great representatives were Hans Georg Gadamer and Michel Foucault, did not limit itself to the interpretation of the text but developed into a much larger and more fundamental trend of philosophical thought. In summary, this trend treated the world as a universal and specific text that carried the message about its existence. Hermeneutics simultaneously decreed the text as an autonomous and significant event. However, no philosopher of this trend thought that it was possible for a fundamentally new instance of text to emerge, which is an artificial text. Descartes' approach was rather obligatory. He firmly denied such competences to a machine, protecting the uniqueness of a human being. At stake was the ability to understand and explore the world, and ultimately to rule it.

R. Maciag, *Knowledge as a Tale*, https://doi.org/10.1007/978-3-031-69820-0

There is an important circumstance that cannot be ignored and is even disturbing due to its persistence. The domain of language appeared again as a practical field in which the ability to produce texts that necessarily contain (hermeneutical) knowledge occurred. Large language models are trained (generated, built) by processing large sets of text which arise as an unintentional product of millions of spontaneous situations of all kinds. The Internet is a huge collection of such texts and can be treated as an unprecedented and incomparable resource of this type. Without the Internet, the creation of large language models would not be possible. On the other hand, the experience of large language models forces us to look at it in a new way: as a reservoir of texts that involuntarily accumulate knowledge. In this situation, the question of what this knowledge is, especially in the context of the growing number of practical applications and the field in which business excels, is justified and urgent.

Observation of the technique of processing large text resources, which governs the creation of the language model, brings some facts, but also new questions. This technique is based on examining and recording the probability that individual, unique words appear next to other words. A sufficiently large set of such situations, which allows one to formulate statistically correct conclusions is needed to calculate this probability. Hence the need to have large resources of texts. This shockingly simple principle, resulting from the so-called distributional hypothesis, is applied in the model algorithm at the beginning and is repeated in more subtle forms. The mentioned probability is recorded in an artificial neural network, thanks to a very ingenious technique. This recording, which is the result of training this network, has a typical shape of neural networks, i.e. is a set of numbers—weights—that determine the behavior of the network. Each tested word receives its own set of such numbers, which contains the mentioned probability.

Writing words as sets of numbers is a fundamental achievement because these sets also turn out to be a record of the semantic relationships of words with other words. The numbers themselves have no interpretation, but their sets turn out to be a mathematical form of mutual semantic relations. These relationships make the text meaningful. A mathematical form can be transformed into a geometric representation. Sets of numbers can be treated as coordinates and create a space in which individual words find their location, reflecting their meaning in the context of other words.

The described mechanism is crucial and is accompanied by a number of additional innovations. For example, structures slightly smaller than words, called tokens, are used as effective carriers of meaning. There is no doubt, however, that it is possible to formalize the semantics of the language, that is, to write it down in the form of numbers which in subsequent steps are subject to various types of calculations. These calculations have a direct semantic interpretation. In other words, semantic computations become possible. These calculations, however, require enormous computing power, which has become practically available, but remains very expensive. Without it, great models could not function. In this context, the economic background of their commonness becomes critical. This background generates questions about the hegemony of the great owners of this technology, who also maintain their dominant position consistently as owners of access to the semantic content of language and knowledge. The paradoxical spread of large language model technologies as consumer products only increases doubts and threats.

The great language models confirm one of the most important premises. Language can create forms capable of absorbing, storing, processing, and articulating knowledge. This language capacity manifests itself through texts. The experience of large language models proves that this mechanism is attainable by machines. This new turn of events in the story of knowledge promises exciting and dramatic events.

Printed in the USA
CPSIA information can be obtained
at www.ICGtesting.com
CBHW050236091124
17166CB00003B/35

9 783031 698194